T0246409

Seed Science and Technology

Seed Science and Technology

Edited by
Percival Rooser

Seed Science and Technology
Edited by Percival Rooser
ISBN: 978-1-63549-255-2 (Hardback)

Published by Larsen and Keller Education,
5 Penn Plaza,
19th Floor,
New York, NY 10001, USA

Cataloging-in-Publication Data

Seed science and technology / edited by Percival Rooser.
p. cm.
Includes bibliographical references and index.
ISBN 978-1-63549-255-2
1. Seeds. 2. Seed technology. 3. Seeds--Development.
I. Rooser, Percival.
SB117 .S44 2017
631.521--dc23

Printed and bound in the United States of America.

For more information regarding Larsen and Keller Education and its products, please visit the publisher's website www.larsen-keller.com

Table of Contents

	Preface	**VII**
Chapter 1	**Introduction to Seed**	**1**
Chapter 2	**Seedling and Germination: An Overview**	**22**
	a. Seedling	22
	b. Germination	31
	c. Epigeal Germination	37
	d. Hypogeal Germination	38
	e. Lily Seed Germination Types	39
	f. Sprouting	40
	g. Sprouted Bread	49
	h. Seed Germinator	51
	i. Seed Dormancy	51
	j. Seed Dispersal	56
Chapter 3	**Essential Elements of Seed Science**	**63**
	a. Seed Production and Gene Diversity	63
	b. Seed Contamination	64
	c. Seed Enhancement	65
	d. Seed Saving	65
	e. Seed Paper	67
	f. Seed Orchard	67
	g. Seed Testing	70
	h. Seed Drill	70
Chapter 4	**Seed Bank: An Integrated Study**	**75**
	a. Seed Bank	75
	b. Soil Seed Bank	78
	c. Svalbard Global Seed Vault	80
	d. Millennium Seed Bank Partnership	85
Chapter 5	**Genetic Technology in Seed Science**	**89**
	a. Genetically Modified Crops	89
	b. Genetically Modified Soybean	106
	c. Genetically Modified Maize	109
Chapter 6	**Seed Producing Plants**	**116**
	a. Embryophyte	116
	b. Spermatophyte	123
	c. Ovule	125
	d. Gymnosperm	130
	e. Flowering Plant	133

Chapter 7	**Pollination: A Comprehensive Study**	**160**
	a. Pollination	160
	b. Open Pollination	181
	c. Pollenizer	182
	d. Self-pollination	182
	e. Pollinator	186
	f. Pollination Trap	192
	g. Pollen Tube	193
	h. Pollination Syndrome	195
	i. Seed Dispersal Syndrome	201
	j. Ornithophily	206
	k. Buzz Pollination	209
	l. Fruit Tree Pollination	210
Chapter 8	**Plant Reproduction**	**215**
	a. Plant Reproduction	215
	b. Certation	221
	c. Diaspore (Botany)	221
	d. Flower	222
	e. Gynoecium	238
	f. Orbicule	245
	g. Petal	246
	h. Stamen	251
	i. Stigma (Botany)	255
	Permissions	
	Index	

Preface

This book is a compilation of chapters that discuss the most vital concepts in the field of seed science and technology. The topics introduced in it cover the basic and primary techniques and methods of the subject. Seed science covers all aspects of the germination of seeds and the related technologies and methodologies that facilitate plant growth. Different approaches, evaluations and methodologies on the subject have been included in this text. For all those who are interested in seed science and technology, this textbook can prove to be an essential guide. This textbook is a complete source of knowledge on the present status of this important field.

To facilitate a deeper understanding of the contents of this book a short introduction of every chapter is written below:

Chapter 1- Seed is a botanical component that causes reproduction in plants. It is the product of the ripened ovule, once it has been fertilized by pollen and there is some growth within the mother plant. This chapter will provide an integrated understanding of seed.

Chapter 2- Seedling is any young plant that is developing from a seed. It typically consists of three main parts, the radicle, the hypocotyl and the cotyledons. Germination is general process of growth of a plant from a seed. This section is an overview of the subject matter incorporating all the major aspects of seedling and germination.

Chapter 3- Seed contamination is the process in which seeds are mixed for the purpose of agriculture. For example, mixing corn seed with the seed of weed. The alternative elements of seed science are seed production and gene diversity, seed enhancement, seed paper, seed testing, speed limit enforcement, seed drill etc. The topics discussed in the chapter are of great importance to broaden the existing knowledge on seed science.

Chapter 4- Seed banks are banks that store seeds; it is done in order to preserve genetic diversity of seeds. Soil seed bank and Svalbard global seed vault are some of the topics discussed in this section. This section will provide an integrated understanding of seed bank.

Chapter 5- Genetically modified crops are crops that have been altered with the help of genetic engineering techniques. It is mainly done with the purpose of adding a trait to the plant. Genetically modified soybean and genetically modified maize have also been explained in the section. The topics discussed in the chapter are of great importance to broaden the existing knowledge on seed science.

Chapter 6- Embryophyte is the most common group of plants that consist the vegetation on Earth. Embryophytes include liverworts, mosses, flowering plants and lycophytes. They are also referred to land plants majorly because of they live in terrestrial habitats. The section serves as a source to understand the main seeds that produce plants.

Chapter 7- Pollination is the process in which pollen is transferred to the female reproductive organ. This process enables fertilization to take place. The forms of pollination elucidated within the section are anemophily, hydrophily, entomophily and zoophily. The chapter strategically encompasses and incorporates the major components of pollination, providing a complete understanding.

Chapter 8- The process of reproduction that takes place in plants is known as plant reproduction. Certation, diaspore, flower, gynoecium, petal, stamen and stigma are some of the aspects that have been explained in the section. The aspects elucidated in this section are of vital importance, and provide a better understanding of plant reproduction system.

Finally, I would like to thank the entire team involved in the inception of this book for their valuable time and contribution. This book would not have been possible without their efforts. I would also like to thank my friends and family for their constant support.

Editor

1

Introduction to Seed

Seed is a botanical component that causes reproduction in plants. It is the product of the ripened ovule, once it has been fertilized by pollen and there is some growth within the mother plant. This chapter will provide an integrated understanding of seed.

Brown flax seeds

A seed is an embryonic plant enclosed in a protective outer covering. The formation of the seed is part of the process of reproduction in seed plants, the spermatophytes, including the gymnosperm and angiosperm plants.

Seeds are the product of the ripened ovule, after fertilization by pollen and some growth within the mother plant. The embryo is developed from the zygote and the seed coat from the integuments of the ovule.

Seeds have been an important development in the reproduction and success of gymnosperms and angiosperms plants, relative to more primitive plants such as ferns, mosses and liverworts, which do not have seeds and use other means to propagate themselves. Seed plants now dominate biological niches on land, from forests to grasslands both in hot and cold climates.

The term "seed" also has a general meaning that antedates the above—anything that can be sown, e.g. "seed" potatoes, "seeds" of corn or sunflower "seeds". In the case of sunflower and corn "seeds", what is sown is the seed enclosed in a shell or husk, whereas the potato is a tuber.

Many structures commonly referred to as "seeds" are actually dry fruits. Plants producing berries are called baccate. Sunflower seeds are sometimes sold commercially while still enclosed within the hard wall of the fruit, which must be split open to reach the seed. Different groups of plants have other modifications, the so-called stone fruits (such as the peach) have a hardened fruit layer (the endocarp) fused to and surrounding the actual seed. Nuts are the one-seeded, hard-shelled fruit of some plants with an indehiscent seed, such as an acorn or hazelnut.

Production

Seeds are produced in several related groups of plants, and their manner of production distinguishes the angiosperms ("enclosed seeds") from the gymnosperms ("naked seeds"). Angiosperm seeds are produced in a hard or fleshy structure called a fruit that encloses the seeds, hence the name. Some fruits have layers of both hard and fleshy material. In gymnosperms, no special structure develops to enclose the seeds, which begin their development "naked" on the bracts of cones. However, the seeds do become covered by the cone scales as they develop in some species of conifer.

Seed production in natural plant populations varies widely from year-to-year in response to weather variables, insects and diseases, and internal cycles within the plants themselves. Over a 20-year period, for example, forests composed of loblolly pine and shortleaf pine produced from 0 to nearly 5 million sound pine seeds per hectare. Over this period, there were six bumper, five poor, and nine good seed crops, when evaluated in regard to producing adequate seedlings for natural forest reproduction.

Development

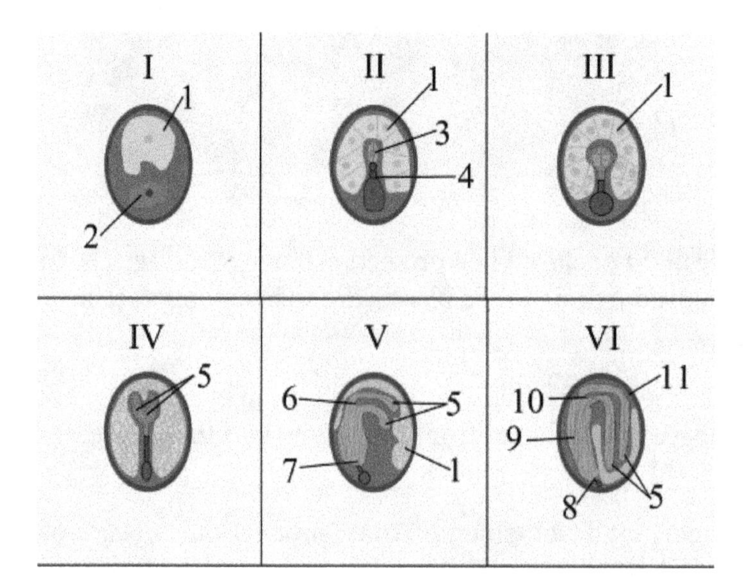

Stages of seed development:

I Zygote	**IV** Heart
II Proembryo	**V** Torpedo
III Globular	**VI** Mature Embryo

Key: *1. Endosperm 2. Zygote 3. Embryo 4. Suspensor 5. Cotyledons 6. Shoot Apical Meristem 7. Root Apical Meristem 8. Radicle 9. Hypocotyl 10. Epicotyl 11. Seed Coat*

Angiosperm (flowering plants) seeds consist of three genetically distinct constituents: (1) the embryo formed from the zygote, (2) the endosperm, which is normally triploid, (3) the seed coat from tissue derived from the maternal tissue of the ovule. In angiosperms, the process of seed development begins with double fertilization, which involves the fusion of two male gametes with the egg cell and the central cell to form the primary endosperm and the zygote. Right after fertilization, the zygote is mostly inactive, but the primary endosperm divides rapidly to form the endosperm tissue. This tissue becomes the food the young plant will consume until the roots have developed after germination.

Ovule

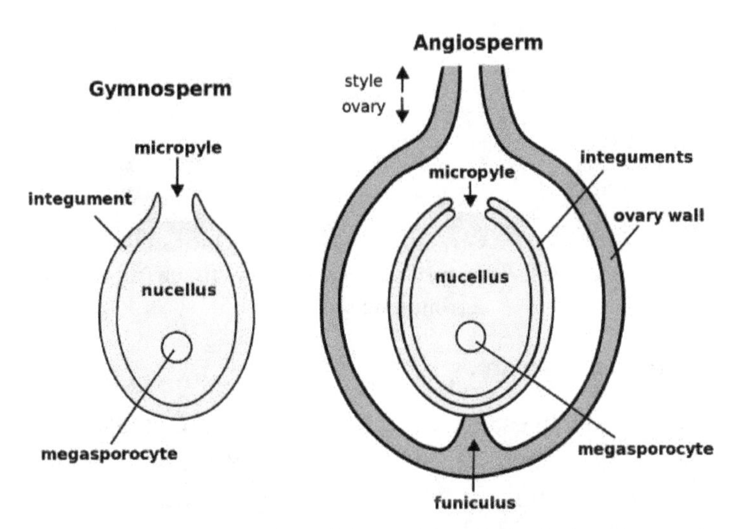

Plant ovules: Gymnosperm ovule on left, angiosperm ovule (inside ovary) on right

After fertilization the ovules develop into the seeds. The ovule consists of a number of components:

- The funicle (*funiculus, funiculi*) or seed stalk which attaches the ovule to the placenta and hence ovary or fruit wall, at the pericarp.
- The nucellus, the remnant of the megasporangium and main region of the ovule where the megagametophyte develops.
- The micropyle, a small pore or opening in the apex of the integument of the ovule where the pollen tube usually enters during the process of fertilization.
- The chalaza, the base of the ovule opposite the micropyle, where integument and nucellus are joined together).

The shape of the ovules as they develop often affects the final shape of the seeds. Plants generally produce ovules of four shapes: the most common shape is called anatropous, with a curved shape. Orthotropous ovules are straight with all the parts of the ovule lined up in a long row producing an uncurved seed. Campylotropous ovules have a curved megagametophyte often giving the seed a tight "C" shape. The last ovule shape is called amphitropous, where the ovule is partly inverted and turned back 90 degrees on its stalk (the funicle or funiculus).

In the majority of flowering plants, the zygote's first division is transversely oriented in regards to the long axis, and this establishes the polarity of the embryo. The upper or chalazal pole becomes the main area of growth of the embryo, while the lower or micropylar pole produces the stalk-like suspensor that attaches to the micropyle. The suspensor absorbs and manufacturers nutrients from the endosperm that are used during the embryo's growth.

Embryo

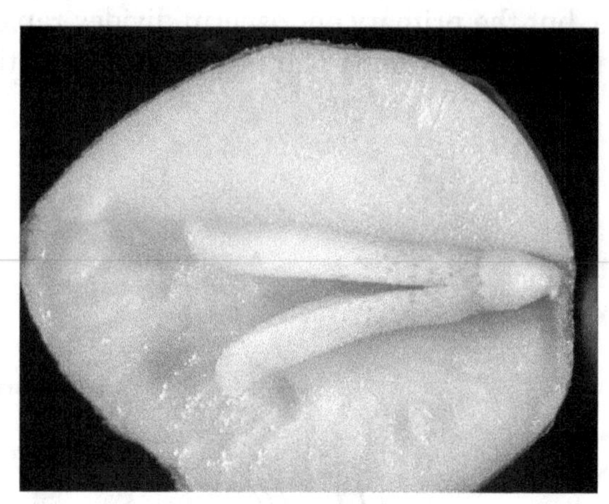

The inside of a *Ginkgo* seed, showing a well-developed embryo, nutritive tissue (megagametophyte), and a bit of the surrounding seed coat

The main components of the embryo are:

- The cotyledons, the seed leaves, attached to the embryonic axis. There may be one (Monocotyledons), or two (Dicotyledons). The cotyledons are also the source of nutrients in the non-endospermic dicotyledons, in which case they replace the endosperm, and are thick and leathery. In endospermic seeds the cotyledons are thin and papery. Dicotyledons have the point of attachment opposite one another on the axis.
- The epicotyl, the embryonic axis above the point of attachment of the cotyledon(s).
- The plumule, the tip of the epicotyl, and has a feathery appearance due to the presence of young leaf primordia at the apex, and will become the shoot upon germination.
- The hypocotyl, the embryonic axis below the point of attachment of the cotyledon(s), connecting the epicotyle and the radicle, being the stem-root transition zone.
- The radicle, the basal tip of the hypocotyl, grows into the primary root.

Monocotyledonous plants have two additional structures in the form of sheaths. The plumule is covered with a coleoptile that forms the first leaf while the radicle is covered with a coleorhiza that connects to the primary root and adventitious roots form from the sides. Here the hypocotyl is a rudimentary axis between radicle and plumule. The seeds of corn are constructed with these structures; pericarp, scutellum (single large cotyledon) that absorbs nutrients from the endosperm, plumule, radicle, coleoptile and coleorhiza–these last two structures are sheath-like and enclose the plumule and radicle, acting as a protective covering.

Seed Coat

The maturing ovule undergoes marked changes in the integuments, generally a reduction and disorganisation but occasionally a thickening. The seed coat forms from the two integuments or outer layers of cells of the ovule, which derive from tissue from the mother plant, the inner integument forms the tegmen and the outer forms the testa. (The seed coats of some mononocotyledon plants, such as the grasses, are not distinct structures, but are fused with the fruit wall to form a pericarp.)

The testae of both monocots and dicots are often marked with patterns and textured markings, or have wings or tufts of hair. When the seed coat forms from only one layer, it is also called the testa, though not all such testae are homologous from one species to the next. The funiculus abscises (detaches at fixed point – abscission zone), the scar forming an oval depression, the hilum. Anatropous ovules have a portion of the funiculus that is adnate (fused to the seed coat), and which forms a longitudinal ridge, or raphe, just above the hilum. In bitegmic ovules (e.g. *Gossypium* described here) both inner and outer integuments contribute to the seed coat formation. With continuing maturation the cells enlarge in the outer integument. While the inner epidermis may remain a single layer, it may also divide to produce two to three layers and accumulates starch, and is referred to as the colourless layer. By contrast the outer epidermis becomes tanniferous. The inner integument may consist of eight to fifteen layers. (Kozlowski 1972)

As the cells enlarge, and starch is deposited in the outer layers of the pigmented zone below the outer epidermis, this zone begins to lignify, while the cells of the outer epidermis enlarge radially and their walls thicken, with nucleus and cytoplasm compressed into the outer layer. these cells which are broader on their inner surface are called palisade cells. In the inner epidermis the cells also enlarge radially with plate like thickening of the walls. The mature inner integument has a palisade layer, a pigmented zone with 15-20 layers, while the innermost layer is known as the fringe layer. (Kozlowski 1972)

Gymnosperms

In gymnosperms, which do not form ovaries, the ovules and hence the seeds are exposed. This is the basis for their nomenclature – naked seeded plants. Two sperm cells transferred from the pollen do not develop the seed by double fertilization, but one sperm nucleus unites with the egg nucleus and the other sperm is not used.

Sometimes each sperm fertilizes an egg cell and one zygote is then aborted or absorbed during early development. The seed is composed of the embryo (the result of fertilization) and tissue from the mother plant, which also form a cone around the seed in coniferous plants such as pine and spruce.

Shape and Appearance

A large number of terms are used to describe seed shapes, many of which are largely self-explanatory such as *Bean-shaped* (reniform) – resembling a kidney, with lobed ends on either side of the hilum, *Square* or *Oblong* – angular with all sides more or less equal or longer than wide, *Triangular* – three sided, broadest below middle, *Elliptic* or *Ovate* or *Obovate* – rounded at both ends, or egg shaped (ovate or obovate, broader at one end), being rounded but either symmetrical about the middle or broader below the middle or broader above the middle.

Other less obvious terms include discoid (resembling a disc or plate, having both thickness and parallel faces and with a rounded margin), ellipsoid, globose (spherical), or subglobose (Inflated, but less than spherical), lenticular, oblong, ovoid, reniform and sectoroid. Striate seeds are striped with parallel, longitudinal lines or ridges. The commonest colours are brown and black, other colours are infrequent. The surface varies from highly polished to considerably roughened. The surface may have a variety of appendages. A seed coat with the consistency of cork is referred to as suberose. Other terms include crustaceous (hard, thin or brittle).

Structure

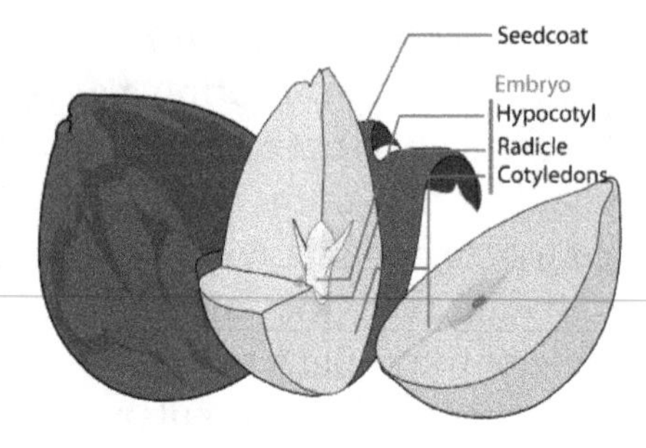

The parts of an avocado seed (a dicot), showing the seed coat and embryo

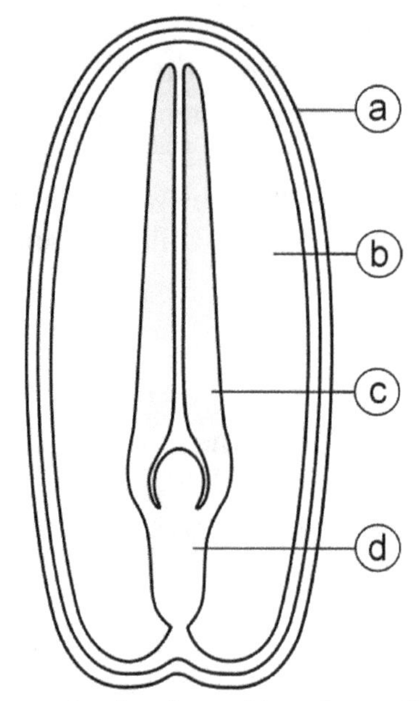

Diagram of the internal structure of a dicot seed and embryo: (a) seed coat, (b) endosperm, (c) cotyledon, (d) hypocotyl

A typical seed includes two basic parts:

1. an embryo;
2. a seed coat.

In addition, the endosperm forms a supply of nutrients for the embryo in most monocotyledons and the endospermic dicotyledons.

Seed Types

Seeds have been considered to occur in many structurally different types (Martin 1946). These are based on a number of criteria, of which the dominant one is the embryo-to-seed size ratio. This reflects the degree to which the developing cotyledons absorb the nutrients of the endosperm, and thus obliterate it.

Six types occur amongst the monocotyledons, ten in the dicotyledons, and two in the gymnosperms (linear and spatulate). This classification is based on three characteristics: embryo morphology, amount of endosperm and the position of the embryo relative to the endosperm.

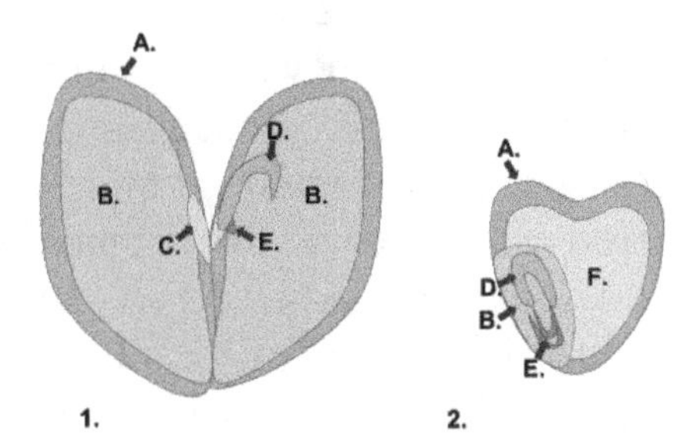

Diagram of a generalized dicot seed (1) versus a generalized monocot seed (2). A. Seed Coat B. Cotyledon C. Hilum D. Plumule E. Radicle F. Endosperm

Embryo

In endospermic seeds, there are two distinct regions inside the seed coat, an upper and larger endosperm and a lower smaller embryo. The embryo is the fertilised ovule, an immature plant from which a new plant will grow under proper conditions. The embryo has one cotyledon or seed leaf in monocotyledons, two cotyledons in almost all dicotyledons and two or more in gymnosperms. In the fruit of grains (caryopses) the single monocotyledon is shield shaped and hence called a scutellum. The scutellum is pressed closely against the endosperm from which it absorbs food, and passes it to the growing parts. Embryo descriptors include small, straight, bent, curved and curled.

Nutrient Storage

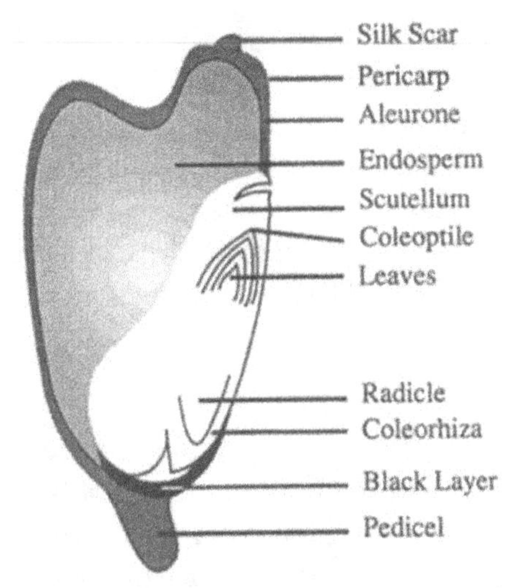

Layers within an endospermic maize seed

Within the seed, there usually is a store of nutrients for the seedling that will grow from the embryo. The form of the stored nutrition varies depending on the kind of plant. In angiosperms, the stored food begins as a tissue called the endosperm, which is derived from the mother plant and the pollen via double fertilization. It is usually triploid, and is rich in oil or starch, and protein. In gymnosperms, such as conifers, the food storage tissue (also called endosperm) is part of the female gametophyte, a haploid tissue. The endosperm is surrounded by the aleurone layer (peripheral endosperm), filled with proteinaceous aleurone grains.

Originally, by analogy with the animal ovum, the outer nucellus layer (perisperm) was referred to as albumen, and the inner endosperm layer as vitellus. Although misleading, the term began to be applied to all the nutrient matter. This terminology persists in referring to endospermic seeds as 'albuminous'. The nature of this material is used in both describing and classifying seeds, in addition to the embryo to endosperm size ratio. The endosperm may be considered to be farinaceous (or mealy) in which the cells are filled with starch, as for instance cereal grains, or not (non-farinaceous). The endosperm may also be referred to as 'fleshy' or 'cartilaginous' with thicker soft cells such as coconut, but may also be oily as in *Ricinus* (castor oil), *Croton* and Poppy. The endosperm is called 'horny' when the cell walls are thicker such as date and coffee, or 'ruminated' if mottled, as in nutmeg, palms and Annonaceae.

In most monocotyledons (such as grasses and palms) and some (endospermic or albuminous) dicotyledons (such as castor beans) the embryo is embedded in the endosperm (and nucellus, which the seedling will use upon germination. In the non-endospermic dicotyledons the endosperm is absorbed by the embryo as the latter grows within the developing seed, and the cotyledons of the embryo become filled with stored food. At maturity, seeds of these species have no endosperm and are also referred to as exalbuminous seeds. The exalbuminous seeds include the legumes (such as beans and peas), trees such as the oak and walnut, vegetables such as squash and radish, and sunflowers. According to Bewley and Black (1978), Brazil nut storage is in hypocotyl, this place of storage is uncommon among seeds. All gymnosperm seeds are albuminous.

Seed Coat

The seed coat develops from the maternal tissue, the integuments, originally surrounding the ovule. The seed coat in the mature seed can be a paper-thin layer (e.g. peanut) or something more substantial (e.g. thick and hard in honey locust and coconut), or fleshy as in the sarcotesta of pomegranate. The seed coat helps protect the embryo from mechanical injury, predators and drying out. Depending on its development, the seed coat is either bitegmic or unitegmic. Bitegmic seeds form a testa from the outer integument and a tegmen from the inner integument while unitegmic seeds have only one integument. Usually parts of the testa or tegmen form a hard protective mechanical layer. The mechanical layer may prevent water penetration and germination. Amongst the barriers may be the presence of lignified sclereids.

The outer integument has a number of layers, generally between four and eight organised into three layers: (a) outer epidermis, (b) outer pigmented zone of two to five layers containing tannin and starch, and (c) inner epidermis. The endotegmen is derived from the inner epidermis of the inner integument, the exotegmen from the outer surface of the inner integument. The endotesta is derived from the inner epidermis of the outer integument, and the outer layer of the testa from the outer surface of the outer integument is referred to as the exotesta. If the exotesta is also the mechanical layer, this is called an exotestal seed, but if the mechanical layer is the endotegmen, then the seed is endotestal. The exotesta may consist of one or more rows of cells that are elongated and pallisade like (e.g. Fabaceae), hence 'palisade exotesta'.

In addition to the three basic seed parts, some seeds have an appendage, an aril, a fleshy outgrowth of the funicle (funiculus), (as in yew and nutmeg) or an oily appendage, an elaiosome (as in *Corydalis*), or hairs (trichomes). In the latter example these hairs are the source of the textile crop cotton. Other seed appendages include the raphe (a ridge), wings, caruncles (a soft spongy outgrowth from the outer integument in the vicinity of the micropyle), spines, or tubercles.

A scar also may remain on the seed coat, called the hilum, where the seed was attached to the ovary wall by the funicle. Just below it is a small pore, representing the micropyle of the ovule.

Size and Seed Set

A collection of various vegetable and herb seeds

Seeds are very diverse in size. The dust-like orchid seeds are the smallest, with about one million seeds per gram; they are often embryonic seeds with immature embryos and no significant energy reserves. Orchids and a few other groups of plants are mycoheterotrophs which depend on mycorrhizal fungi for nutrition during germination and the early growth of the seedling. Some terrestrial orchid seedlings, in fact, spend the first few years of their lives deriving energy from the fungi and do not produce green leaves. At over 20 kg, the largest seed is the *coco de mer*. Plants that produce smaller seeds can generate many more seeds per flower, while plants with larger seeds invest more resources into those seeds and normally produce fewer seeds. Small seeds are quicker to ripen and can be dispersed sooner, so fall blooming plants often have small seeds. Many annual plants produce great quantities of smaller seeds; this helps to ensure at least a few will end in a favorable place for growth. Herbaceous perennials and woody plants often have larger seeds; they can produce seeds over many years, and larger seeds have more energy reserves for germination and seedling growth and produce larger, more established seedlings after germination.

Functions

Seeds serve several functions for the plants that produce them. Key among these functions are nourishment of the embryo, dispersal to a new location, and dormancy during unfavorable conditions. Seeds fundamentally are means of reproduction, and most seeds are the product of sexual reproduction which produces a remixing of genetic material and phenotype variability on which natural selection acts.

Embryo Nourishment

Seeds protect and nourish the embryo or young plant. They usually give a seedling a faster start than a sporeling from a spore, because of the larger food reserves in the seed and the multicellularity of the enclosed embryo.

Dispersal

Unlike animals, plants are limited in their ability to seek out favorable conditions for life and growth. As a result, plants have evolved many ways to disperse their offspring by dispersing their seeds. A seed must somehow "arrive" at a location and be there at a time favorable for germination and growth. When the fruits open and release their seeds in a regular way, it is called dehiscent, which is often distinctive for related groups of plants; these fruits include capsules, follicles, legumes, silicles and siliques. When fruits do not open and release their seeds in a regular fashion, they are called indehiscent, which include the fruits achenes, caryopsis, nuts, samaras, and utricles.

By Wind (Anemochory)

Dandelion seeds are contained within achenes, which can be carried long distances by the wind.

The seed pod of milkweed (*Asclepias syriaca*)

- Some seeds (e.g., pine) have a wing that aids in wind dispersal.
- The dustlike seeds of orchids are carried efficiently by the wind.
- Some seeds (e.g. milkweed, poplar) have hairs that aid in wind dispersal.

Other seeds are enclosed in fruit structures that aid wind dispersal in similar ways:

- Dandelion achenes have hairs.
- Maple samaras have two wings.

By Water (Hydrochory)

- Some plants, such as *Mucuna* and *Dioclea*, produce buoyant seeds termed sea-beans or drift seeds because they float in rivers to the oceans and wash up on beaches.

By Animals (Zoochory)

- Seeds (burrs) with barbs or hooks (e.g. acaena, burdock, dock) which attach to animal fur or feathers, and then drop off later.
- Seeds with a fleshy covering (e.g. apple, cherry, juniper) are eaten by animals (birds, mammals, reptiles, fish) which then disperse these seeds in their droppings.
- Seeds (nuts) are attractive long-term storable food resources for animals (e.g. acorns, hazelnut, walnut); the seeds are stored some distance from the parent plant, and some escape being eaten if the animal forgets them.

Myrmecochory is the dispersal of seeds by ants. Foraging ants disperse seeds which have appendages called elaiosomes (e.g. bloodroot, trilliums, acacias, and many species of Proteaceae). Elaiosomes are soft, fleshy structures that contain nutrients for animals that eat them. The ants carry such seeds back to their nest, where the elaiosomes are eaten. The remainder of the seed, which is hard and inedible to the ants, then germinates either within the nest or at a removal site where the seed has been discarded by the ants. This dispersal relationship is an example of mutualism, since the plants depend upon the ants to disperse seeds, while the ants depend upon the plants seeds for food. As a result, a drop in numbers of one partner can reduce success of the other. In South Africa, the Argentine ant (*Linepithema humile*) has invaded and displaced native species of ants. Unlike the native ant species, Argentine ants do not collect the seeds of *Mimetes cucullatus* or eat the elaiosomes. In areas where these ants have invaded, the numbers of *Mimetes* seedlings have dropped.

Dormancy

Seed dormancy has two main functions: the first is synchronizing germination with the optimal conditions for survival of the resulting seedling; the second is spreading germination of a batch of seeds over time so a catastrophe (e.g. late frosts, drought, herbivory) does not result in the death of all offspring of a plant (bet-hedging). Seed dormancy is defined as a seed failing to germinate under environmental conditions optimal for germination, normally when the environment is at a suitable temperature with proper soil moisture. This true dormancy or innate dormancy is therefore caused by conditions within the seed that prevent germination. Thus dormancy is a state of the seed, not of the environment. Induced dormancy, enforced dormancy or seed quiescence occurs when a seed fails to germinate because the external environmental conditions are inappropriate for germination, mostly in response to conditions being too dark or light, too cold or hot, or too dry.

Seed dormancy is not the same as seed persistence in the soil or on the plant, though even in scientific publications dormancy and persistence are often confused or used as synonyms.

Often, seed dormancy is divided into four major categories: exogenous; endogenous; combinational; and secondary. A more recent system distinguishes five classes: morphological, physiological, morphophysiological, physical and combinational dormancy.

Exogenous dormancy is caused by conditions outside the embryo, including:

- Physical dormancy or hard seed coats occurs when seeds are impermeable to water. At dormancy break, a specialized structure, the 'water gap', is disrupted in response to environmental cues, especially temperature, so water can enter the seed and germination can occur. Plant families where physical dormancy occurs include Anacardiaceae, Cannaceae, Convulvulaceae, Fabaceae and Malvaceae.
- Chemical dormancy considers species that lack physiological dormancy, but where a chemical prevents germination. This chemical can be leached out of the seed by rainwater or snow melt or be deactivated somehow. Leaching of chemical inhibitors from the seed by rain water is often cited as an important cause of dormancy release in seeds of desert plants, but little evidence exists to support this claim.

Endogenous dormancy is caused by conditions within the embryo itself, including:

- In morphological dormancy, germination is prevented due to morphological characteristics of the embryo. In some species, the embryo is just a mass of cells when seeds are dispersed; it is not differentiated. Before germination can take place, both differentiation and growth of the embryo have to occur. In other species, the embryo is differentiated but not fully grown (underdeveloped) at dispersal, and embryo growth up to a species specific length is required before germination can occur. Examples of plant families where morphological dormancy occurs are Apiaceae, Cycadaceae, Liliaceae, Magnoliaceae and Ranunculaceae.
- Morphophysiological dormancy includes seeds with underdeveloped embryos, and also have physiological components to dormancy. These seeds, therefore, require a dormancy-breaking treatments, as well as a period of time to develop fully grown embryos. Plant families where morphophysiological dormancy occurs include Apiaceae, Aquifoliaceae, Liliaceae, Magnoliaceae, Papaveraceae and Ranunculaceae. Some plants with morphophysiological dormancy, such as *Asarum* or *Trillium* species, have multiple types of dormancy, one affects radicle (root) growth, while the other affects plumule (shoot) growth. The terms "double dormancy" and "two-year seeds" are used for species whose seeds need two years to complete germination or at least two winters and one summer. Dormancy of the radicle (seedling root)is broken during the first winter after dispersal while dormancy of the shoot bud is broken during the second winter.

- Physiological dormancy means the embryo, due to physiological causes, cannot generate enough power to break through the seed coat, endosperm or other covering structures. Dormancy is typically broken at cool wet, warm wet or warm dry conditions. Abscisic acid is usually the growth inhibitor in seeds, and its production can be affected by light.
    - Drying, in some plants, including a number of grasses and those from seasonally arid regions, is needed before they will germinate. The seeds are released, but need to have a lower moisture content before germination can begin. If the seeds remain moist after dispersal, germination can be delayed for many months or even years. Many herbaceous plants from temperate climate zones have physiological dormancy that disappears with drying of the seeds. Other species will germinate after dispersal only under very narrow temperature ranges, but as the seeds dry, they are able to germinate over a wider temperature range.
- In seeds with combinational dormancy, the seed or fruit coat is impermeable to water and the embryo has physiological dormancy. Depending on the species, physical dormancy can be broken before or after physiological dormancy is broken.
- Secondary dormancy* is caused by conditions after the seed has been dispersed and occurs in some seeds when nondormant seed is exposed to conditions that are not favorable to germination, very often high temperatures. The mechanisms of secondary dormancy are not yet fully understood, but might involve the loss of sensitivity in receptors in the plasma membrane.

The following types of seed dormancy do not involve seed dormancy, strictly speaking, as lack of germination is prevented by the environment, not by characteristics of the seed itself:

- Photodormancy or light sensitivity affects germination of some seeds. These photoblastic seeds need a period of darkness or light to germinate. In species with thin seed coats, light may be able to penetrate into the dormant embryo. The presence of light or the absence of light may trigger the germination process, inhibiting germination in some seeds buried too deeply or in others not buried in the soil.
- Thermodormancy is seed sensitivity to heat or cold. Some seeds, including cocklebur and amaranth, germinate only at high temperatures (30 °C or 86 °F); many plants that have seeds that germinate in early to midsummer have thermodormancy, so germinate only when the soil temperature is warm. Other seeds need cool soils to germinate, while others, such as celery, are inhibited when soil temperatures are too warm. Often, thermodormancy requirements disappear as the seed ages or dries.

Not all seeds undergo a period of dormancy. Seeds of some mangroves are viviparous; they begin to germinate while still attached to the parent. The large, heavy root allows the seed to penetrate into the ground when it falls. Many garden plant seeds will germinate readily as soon as they have water and are warm enough; though their wild ancestors may have had dormancy, these cultivated plants lack it. After many generations of selective pressure by plant breeders and gardeners, dormancy has been selected out.

For annuals, seeds are a way for the species to survive dry or cold seasons. Ephemeral plants are usually annuals that can go from seed to seed in as few as six weeks.

Persistence and Seed Banks

Germination

Germinating sunflower seedlings

Seed germination is a process by which a seed embryo develops into a seedling. It involves the reactivation of the metabolic pathways that lead to growth and the emergence of the radicle or seed root and plumule or shoot. The emergence of the seedling above the soil surface is the next phase of the plant's growth and is called seedling establishment.

Three fundamental conditions must exist before germination can occur. (1) The embryo must be alive, called seed viability. (2) Any dormancy requirements that prevent germination must be overcome. (3) The proper environmental conditions must exist for germination.

Seed viability is the ability of the embryo to germinate and is affected by a number of different conditions. Some plants do not produce seeds that have functional complete embryos, or the seed may have no embryo at all, often called empty seeds. Predators and pathogens can damage or kill the seed while it is still in the fruit or after it is dispersed. Environmental conditions like flooding or heat can kill the seed before or during germination. The age of the seed affects its health and germination ability: since the seed has a living embryo, over time cells die and cannot be replaced. Some seeds can live for a long time before germination, while others can only survive for a short period after dispersal before they die.

Seed vigor is a measure of the quality of seed, and involves the viability of the seed, the germination percentage, germination rate and the strength of the seedlings produced.

The germination percentage is simply the proportion of seeds that germinate from all seeds subject to the right conditions for growth. The germination rate is the length of time it takes for the seeds to germinate. Germination percentages and rates are affected by seed viability, dormancy and environmental effects that impact on the seed and seedling. In agriculture and horticulture quality seeds have high viability, measured by germination percentage plus the rate of germination. This is given as a percent of germination over a certain amount of time, 90% germination in 20 days, for example. 'Dormancy' is covered above; many plants produce seeds with varying degrees of dormancy, and different seeds from the same fruit can have different degrees of dormancy. It's possible to have seeds with no dormancy if they are dispersed right away and do not dry (if the seeds dry they go into physiological dormancy). There is great variation amongst plants and a dormant seed is still a viable seed even though the germination rate might be very low.

Environmental conditions affecting seed germination include; water, oxygen, temperature and light.

Three distinct phases of seed germination occur: water imbibition; lag phase; and radicle emergence.

In order for the seed coat to split, the embryo must imbibe (soak up water), which causes it to swell, splitting the seed coat. However, the nature of the seed coat determines how rapidly water can penetrate and subsequently initiate germination. The rate of imbibition is dependent on the permeability of the seed coat, amount of water in the environment and the area of contact the seed has to the source of water. For some seeds, imbibing too much water too quickly can kill the seed. For some seeds, once water is imbibed the germination process cannot be stopped, and drying then becomes fatal. Other seeds can imbibe and lose water a few times without causing ill effects, but drying can cause secondary dormancy.

Repair of DNA Damage

During seed dormancy, often associated with unpredictable and stressful environments, DNA damages accumulate as the seeds age. In rye seeds, the reduction of DNA integrity due to damage is associated with loss of seed viability during storage. Upon germination, seeds of *Vicia faba* undergo DNA repair. A plant DNA ligase that is involved in repair of single- and double-strand breaks during seed germination is an important determinant of seed longevity. Also, in Arabidopsis seeds, the activities of the DNA repair enzymes Poly ADP ribose polymerases (PARP) are likely needed for successful germination. Thus DNA damages that accumulate during dormancy appear to be a problem for seed survival, and the enzymatic repair of DNA damages during germination appears to be important for seed viability.

Inducing Germination

A number of different strategies are used by gardeners and horticulturists to break seed dormancy.

Scarification allows water and gases to penetrate into the seed; it includes methods to physically break the hard seed coats or soften them by chemicals, such as soaking in hot water or poking holes in the seed with a pin or rubbing them on sandpaper or cracking with a press or hammer. Sometimes fruits are harvested while the seeds are still immature and the seed coat is not fully developed and sown right away before the seed coat become impermeable. Under natural conditions, seed coats are worn down by rodents chewing on the seed, the seeds rubbing against rocks (seeds are moved by the wind or water currents), by undergoing freezing and thawing of surface water, or passing through an animal's digestive tract. In the latter case, the seed coat protects the seed from digestion, while often weakening the seed coat such that the embryo is ready to sprout when it is deposited, along with a bit of fecal matter that acts as fertilizer, far from the parent plant. Microorganisms are often effective in breaking down hard seed coats and are sometimes used by people as a treatment; the seeds are stored in a moist warm sandy medium for several months under nonsterile conditions.

Stratification, also called moist-chilling, breaks down physiological dormancy, and involves the addition of moisture to the seeds so they absorb water, and they are then subjected to a period of moist chilling to after-ripen the embryo. Sowing in late summer and fall and allowing to overwinter under cool conditions is an effective way to stratify seeds; some seeds respond more favorably to periods of oscillating temperatures which are a part of the natural environment.

Leaching or the soaking in water removes chemical inhibitors in some seeds that prevent germination. Rain and melting snow naturally accomplish this task. For seeds planted in gardens, running water is best—if soaked in a container, 12 to 24 hours of soaking is sufficient. Soaking longer, especially in stagnant water, can result in oxygen starvation and seed death. Seeds with hard seed coats can be soaked in hot water to break open the impermeable cell layers that prevent water intake.

Other methods used to assist in the germination of seeds that have dormancy include prechilling, predrying, daily alternation of temperature, light exposure, potassium nitrate, the use of plant growth regulators, such as gibberellins, cytokinins, ethylene, thiourea, sodium hypochlorite, and others. Some seeds germinate best after a fire. For some seeds, fire cracks hard seed coats, while in others, chemical dormancy is broken in reaction to the presence of smoke. Liquid smoke is often used by gardeners to assist in the germination of these species.

Sterile Seeds

Seeds may be sterile for few reasons: they may have been irradiated, unpollinated, cells lived past expectancy, or bred for the purpose.

Evolution and Origin of Seeds

The origin of seed plants is a problem that still remains unsolved. However, more and more data tends to place this origin in the middle Devonian. The description in 2004 of the proto-seed *Runcaria heinzelinii* in the Givetian of Belgium is an indication of that ancient origin of seed-plants. As with modern ferns, most land plants before this time reproduced by sending spores into the air, that would land and become whole new plants.

The first "true" seeds are described from the upper Devonian, which is probably the theater of their true first evolutionary radiation. The seed plants progressively became one of the major elements of nearly all ecosystems.

Economic Importance

Phaseolus vulgaris seeds are diverse in size, shape, and color.

Edible Seeds

Many seeds are edible and the majority of human calories comes from seeds, especially from cereals, legumes and nuts. Seeds also provide most cooking oils, many beverages and spices and some important food additives. In different seeds the seed embryo or the endosperm dominates and provides most of the nutrients. The storage proteins of the embryo and endosperm differ in their amino acid content and physical properties. For example, the gluten of wheat, important in providing the elastic property to bread dough is strictly an endosperm protein.

Seeds are used to propagate many crops such as cereals, legumes, forest trees, turfgrasses and pasture grasses. Particularly in developing countries, a major constraint faced is the inadequacy of the marketing channels to get the seed to poor farmers. Thus the use of farmer-retained seed remains quite common.

Seeds are also eaten by animals, and are fed to livestock. Many seeds are used as birdseed.

Poison and Food Safety

While some seeds are edible, others are harmful, poisonous or deadly. Plants and seeds often contain chemical compounds to discourage herbivores and seed predators. In some cases, these compounds simply taste bad (such as in mustard), but other compounds are toxic or break down into toxic compounds within the digestive system. Children, being smaller than adults, are more susceptible to poisoning by plants and seeds.

A deadly poison, ricin, comes from seeds of the castor bean. Reported lethal doses are anywhere from two to eight seeds, though only a few deaths have been reported when castor beans have been ingested by animals.

In addition, seeds containing amygdalin—apple, apricot, bitter almond, peach, plum, cherry, quince, and others—when consumed in sufficient amounts, may cause cyanide poisoning. Other seeds that contain poisons include annona, cotton, custard apple, datura, uncooked durian, golden chain, horse-chestnut, larkspur, locoweed, lychee, nectarine, rambutan, rosary pea, sour sop, sugar apple, wisteria, and yew. The seeds of the strychnine tree are also poisonous, containing the poison strychnine.

The seeds of many legumes, including the common bean (*Phaseolus vulgaris*), contain proteins called lectins which can cause gastric distress if the beans are eaten without cooking. The common bean and many others, including the soybean, also contain trypsin inhibitors which interfere with the action of the digestive enzyme trypsin. Normal cooking processes degrade lectins and trypsin inhibitors to harmless forms.

Other Uses

Cotton fiber grows attached to cotton plant seeds. Other seed fibers are from kapok and milkweed.

Many important nonfood oils are extracted from seeds. Linseed oil is used in paints. Oil from jojoba and crambe are similar to whale oil.

Seeds are the source of some medicines including castor oil, tea tree oil and the cancer drug, Laetrile.

Many seeds have been used as beads in necklaces and rosaries including Job's tears, Chinaberry, rosary pea, and castor bean. However, the latter three are also poisonous.

Other seed uses include:

- Seeds once used as weights for balances.
- Seeds used as toys by children, such as for the game Conkers.
- Resin from *Clusia rosea* seeds used to caulk boats.
- Nematicide from milkweed seeds.
- Cottonseed meal used as animal feed and fertilizer.

Seed Records

The massive fruit of the coco de mer

- The oldest viable carbon-14-dated seed that has grown into a plant was a Judean date palm seed about 2,000 years old, recovered from excavations at Herod the Great's palace on Masada in Israel. It was germinated in 2005. (A reported regeneration of *Silene stenophylla* (narrow-leafed campion) from material preserved for 31,800 years in the Siberian permafrost was achieved using fruit tissue, not seed.)
- The largest seed is produced by the coco de mer, or "double coconut palm", *Lodoicea maldivica*. The entire fruit may weigh up to 23 kilograms (50 pounds) and usually contains a single seed.
- The smallest seeds are produced by epiphytic orchids. They are only 85 micrometers long, and weigh 0.81 micrograms. They have no endosperm and contain underdeveloped embryos.
- The earliest fossil seeds are around 365 million years old from the Late Devonian of West Virginia. The seeds are preserved immature ovules of the plant *Elkinsia polymorpha*.

In Religion

The Book of Genesis in the Old Testament begins with an explanation of how all plant forms began:

And God said, Let the earth bring forth grass, the herb yielding seed, and the fruit tree yielding fruit after his kind, whose seed is in itself, upon the earth: and it was so. And the earth brought forth grass, and herb yielding seed after its kind, and the tree yielding fruit, whose seed was in itself, after its kind: and God saw that it was good. And the evening and the morning were the third day.

The Quran speaks about seed germination:

It is Allah Who causeth the seed-grain and the date-stone to split and sprout. He causeth the living to issue from the dead, and He is the one to cause the dead to issue from the living. That is Allah: then how are ye deluded away from the truth?

Hybrid Seed

In agriculture and gardening, hybrid seed is seed produced by cross-pollinated plants. Hybrid seed production is predominant in agriculture and home gardening. It is one of the main contributors to the dramatic rise in agricultural output during the last half of the 20th century. The alternatives to hybridization are open pollination and clonal propagation.

All of the hybrid seeds planted by the farmer will produce similar plants while the seeds of the next generation from those hybrids will not consistently have the desired characteristics. Controlled hybrids provide very uniform characteristics because they are produced by crossing two inbred strains. Elite inbred strains are used that express well-documented and consistent phenotypes (such as high crop yield) that are relatively good for inbred plants.

Hybrids are chosen to improve the characteristics of the resulting plants, such as better yield, greater uniformity, improved color, disease resistance. An important factor is the heterosis or *combining ability* of the parent plants. Crossing any particular pair of inbred strains may or may not result in superior offspring. The parent strains used are therefore carefully chosen so as to achieve the uniformity that comes from the uniformity of the parents, and the superior performance that comes from heterosis.

History

In the US, experimental agriculture stations in the 1920s investigated the hybrid crops, and by the 1930s farmers had widely adopted the first hybrid maize.

Recalcitrant Seed

Photo of one whole and one split mango displaying its seed, which is approximately 1/3 the size of the entire fruit

Recalcitrant seeds (subsequently known as unorthodox seeds) are seeds that do not survive drying and freezing during ex-situ conservation and vice versa. By and large, these seeds cannot resist the effects of drying or temperatures less than 10 °C; thus, they cannot be stored for long periods like orthodox seeds because they can lose their viability. Plants that produce recalcitrant seeds include avocado, mango, mangosteen, lychee, cocoa, rubber tree, some horticultural trees, and several plants used in traditional medicine, such as species of *Virola* and *Pentaclethra*. Generally speaking, most tropical pioneer species have orthodox seeds but many climax species have recalcitrant or intermediate seeds.

Mechanisms of Damage

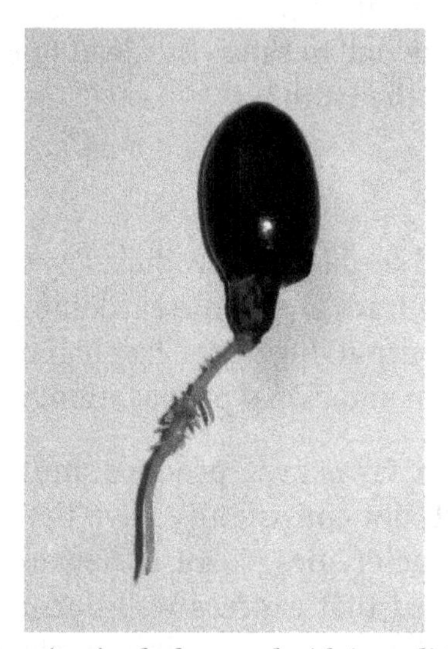

Germinating lychee seed with its radicle

The two main mechanisms of action of damage to recalcitrant seeds are desiccation effect on the intracellular structures and the effect of metabolic damage from the formation of toxic chemicals such as free radicals. An example of the first type of damage would be found in some recalcitrant nontropical hardwood seeds, specifically the acorns of recalcitrant oaks, which can be stored in a nonfrozen state for up to two years provided that precautions be taken against drying. These seeds showed deterioration of cell membrane lipids and proteins after as few as 3–4 days of drying. Other seeds such as those of the sweet chestnut (*Castanea sativa*) show oxidative damage resulting from uncontrolled metabolism occurring during the drying process.

Orthodox Seed

Zea maize, a widely grown orthodox seed which may be dried for two years without harm

Orthodox seeds are seeds which will survive drying and/or freezing during ex-situ conservation. According to information from the U.S. Department of Agriculture, there is variation in the ability of orthodox seeds to withstand drying and storage, with some seeds being more sensitive than others. Thus some seeds are considered intermediate in their storage capability while others are fully orthodox. One notable example of a long-lived orthodox seed which survived accidental storage followed by controlled germination is the case of the 2,000-year-old Judean date palm (cultivar of *Phoenix dactylifera*) seed which successfully sprouted in 2005. This particular seed is reputed to be the oldest viable seed, but the upper survival time limit of properly stored seeds remains unknown.

References

- Galili G; Kigel J (1995). "Chapter One". Seed development and germination. New York: M. Dekker. ISBN 0-8247-9229-7.
- Rost, Thomas L.; Weier, T. Elliot; Weier, Thomas Elliot (1979). Botany: a brief introduction to plant biology. New York: Wiley. p. 319. ISBN 0-471-02114-8.
- "Carol C. Baskin, Jerry M. Baskin. Seeds: Ecology, Biogeography, and Evolution of Dormancy and Germination. Elsevier, 2001". google.ca. p. 27. ISBN 0-12-080263-5.
- "6 – Seed and fruit – University Publishing Online – Paula J. Rudall. Anatomy of Flowering Plants: An Introduction to Structure and Development. Third edition". Cambridge University Press. 2007. ISBN 978-0-521-69245-8.
- Morhardt, Sia; Morhardt, Emil; Emil Morhardt, J. (2004). California desert flowers: an introduction to families, genera, and species. Berkeley: University of California Press. p. 24. ISBN 0-520-24003-0.
- Ricklefs, Robert E. (1993) The Economy of Nature, 3rd ed., p.396. (New York: W. H. Freeman). ISBN 0-7167-2409-X.
- Black, Michael H.; Halmer, Peter (2006). The encyclopedia of seeds: science, technology and uses. Wallingford, UK: CABI. p. 224. ISBN 978-0-85199-723-0.
- Hartmann, Hudson Thomas, and Dale E. Kester. 1983. Plant propagation principles and practices. Englewood Cliffs, N.J.: Prentice-Hall. ISBN 0-13-681007-1. Pages 175-77.
- Taylor EL; Taylor TMC (1993). The biology and evolution of fossil plants. Englewood Cliffs, N.J: Prentice Hall. p. 466. ISBN 0-13-651589-4.
- "Improving Corn; based on "Hybrid Corn", published in the Yearbook of Agriculture, 1962". United States Department of Agriculture. Retrieved 14 December 2014.
- Flores, E.M.; J. A. Vozzo Editor. "Ch 1. Seed Biology" (PDF). Tropical Tree Seed Manual. USDA Forest Service. Retrieved 2011-12-24.
- Berjak, Patricia; N.W. Pammenter; J. A. Vozzo Editor. "Ch 4. Orthodox and Recalcitrant Seeds" (PDF). Tropical Tree Seed Manual. USDA Forest Service. Retrieved 2010-09-27.

2

Seedling and Germination: An Overview

Seedling is any young plant that is developing from a seed. It typically consists of three main parts, the radicle, the hypocotyl and the cotyledons. Germination is general process of growth of a plant from a seed. This section is an overview of the subject matter incorporating all the major aspects of seedling and germination.

Seedling

Monocot]] (left) and dicot (right) seedlings

A seedling is a young plant sporophyte developing out of a plant embryo from a seed. Seedling development starts with germination of the seed. A typical young seedling consists of three main parts: the radicle (embryonic root), the hypocotyl (embryonic shoot), and the cotyledons (seed leaves). The two classes of flowering plants (angiosperms) are distinguished by their numbers of seed leaves: monocotyledons (monocots) have one blade-shaped cotyledon, whereas dicotyledons (dicots) possess two round cotyledons. Gymnosperms are more varied. For example, pine seedlings have up to eight cotyledons. The seedlings of some flowering plants have no cotyledons at all. These are said to be acotyledons.

The plumule is the part of a seed embryo that develops into the shoot bearing the first true leaves of a plant. In most seeds, for example the sunflower, the plumule is a small conical structure without any leaf structure. Growth of the plumule does not occur until the cotyledons have grown above ground. This is epigeal germination. However, in seeds such as the broad bean, a leaf structure is visible on the plumule in the seed. These seeds develop by the plumule growing up through the soil with the cotyledons remaining below the surface. This is known as hypogeal germination.

Photomorphogenesis and Etiolation

Dicot seedlings grown in the light develop short hypocotyls and open cotyledons exposing the epicotyl. This is also referred to as photomorphogenesis. In contrast, seedlings grown in the dark develop long hypocotyls and their cotyledons remain closed around the epicotyl in an *apical hook.* This is referred to as skotomorphogenesis or etiolation. Etiolated seedlings are yellowish in color as chlorophyll synthesis and chloroplast development depend on light. They will open their cotyledons and turn green when treated with light.

In a natural situation, seedling development starts with skotomorphogenesis while the seedling is growing through the soil and attempting to reach the light as fast as possible. During this phase, the cotyledons are tightly closed and form the *apical hook* to protect the shoot apical meristem from damage while pushing through the soil. In many plants, the seed coat still covers the cotyledons for extra protection.

Upon breaking the surface and reaching the light, the seedling's developmental program is switched to photomorphogenesis. The cotyledons open upon contact with light (splitting the seed coat open, if still present) and become green, forming the first photosynthetic organs of the young plant. Until this stage, the seedling lives off the energy reserves stored in the seed. The opening of the cotyledons exposes the shoot apical meristem and the *plumule* consisting of the first *true leaves* of the young plant.

The seedlings sense light through the light receptors phytochrome (red and far-red light) and cryptochrome (blue light). Mutations in these photo receptors and their signal transduction components lead to seedling development that is at odds with light conditions, for example seedlings that show photomorphogenesis when grown in the dark.

Seedling Growth and Maturation

Seedling of a dicot, *Nandina domestica*, showing two green cotyledon leaves, and the first "true" leaf with its distinct leaflets and red-green color.

Once the seedling starts to photosynthesize, it is no longer dependent on the seed's energy reserves. The apical meristems start growing and give rise to the root and shoot. The first "true" leaves expand and can often be distinguished from the round cotyledons through their species-dependent distinct shapes. While the plant is growing and developing additional leaves, the cotyledons eventually senesce and fall off. Seedling growth is also affected by mechanical stimulation, such as by wind or other forms of physical contact, through a process called thigmomorphogenesis.

Temperature and light intensity interact in their effect on seedling growth; at low light levels about 40 lumens/m^2 a day/night temperature regime of 28 °C/13 °C is effective (Brix 1972). A photoperiod shorter than 14 hours causes growth to stop, whereas a photoperiod extended with low light intensities to 16 h or more brings about continuous (free) growth. Little is gained by using more than 16 h of low light intensity once seedlings are in the free growth mode. Long photoperiods using high light intensities from 10,000 to 20,000 lumens/m^2 increase dry matter production, and increasing the photoperiod from 15 to 24 hours may double dry matter growth (Pollard and Logan 1976, Carlson 1979).

The effects of carbon dioxide enrichment and nitrogen supply on the growth of white spruce and trembling aspen were investigated by Brown and Higginbotham (1986). Seedlings were grown in controlled environments with ambient or enriched atmospheric CO_2 (350 or 750 ﬁ/L, respectively) and with nutrient solutions with high, medium, and low N content (15.5, 1.55, and 0.16 mM). Seedlings were harvested, weighed, and measured at intervals of less than 100 days. N supply strongly affected biomass accumulation, height, and leaf area of both species. In white spruce only, the root weight ratio (RWR) was significantly increased with the low-nitrogen regime. CO_2 enrichment for 100 days significantly increased the leaf and total biomass of white spruce seedlings in the high-N regime, RWR of seedlings in the medium-N regime, and root biomass of seedlings in the low-N regime.

First-year seedlings typically have high mortality rates, drought being the principal cause, with roots having been unable to develop enough to maintain contact with soil sufficiently moist to prevent the development of lethal seedling water stress. Somewhat paradoxically, however, Eis (1967a) observed that on both mineral and litter seedbeds, seedling mortality was greater in moist habitats (alluvium and *Aralia–Dryopteris*) that in dry habitats (*Cornus*–Moss). He commented that in dry habitats after the first growing season surviving seedlings appeared to have a much better chance of continued survival than those in moist or wet habitats, in which frost heave and competition from lesser vegetation became major factors in later years. The annual mortality documented by Eis (1967a) is instructive.

Pests and Diseases

Seedlings are particularly vulnerable to attack by pests and diseases and can consequently experience high mortality rates. Pests and diseases which are especially damaging to seedlings include damping off, cutworms, slugs and snails.

Transplanting

Seedlings are generally transplanted when the first pair of true leaves appear. A shade may be provided if the area is arid or hot. A commercially available vitamin hormone concentrate may be used to avoid transplant shock which may contain thiamine hydrochloride, naphthly acetic acid and indole butyric acid.

Images

A few days old Scots pine seedling, the seed still protecting the cotyledons.

Seedling

Seedling of *Quercus robur* sprouting from its acorn.

Dicotyledon plantlet showing roots.

Parts of Seedling

Radicle

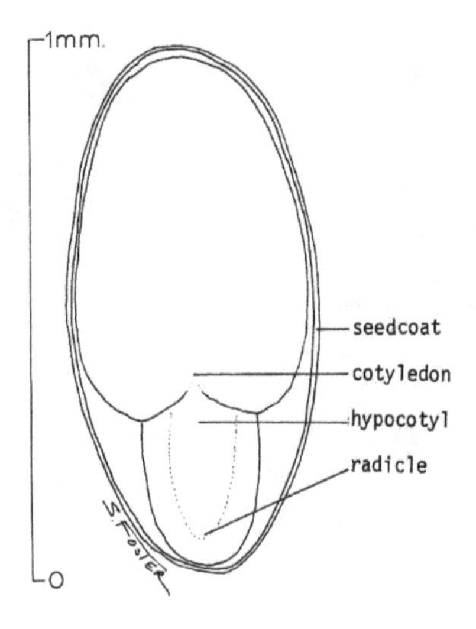

seed of Scouler's willow (*Salix scouleriana*)

In botany, the radicle is the first part of a seedling (a growing plant embryo) to emerge from the seed during the process of germination. The *radicle* is the embryonic root of the plant, and grows downward in the soil (the shoot emerges from the plumule). Above the radicle is the embryonic stem or hypocotyl, supporting the cotyledon(s).

It is the embryonic root inside the seed. It is the first thing to emerge from a seed and down into the ground to allow the seed to suck up water and send out its leaves so that it can start photosynthesizing.

The radicle emerges from a seed through the micropyle. Radicles in seedlings are classified into two main types. Those pointing away from the seed coat scar or hilum are classified as antitropous, and those pointing towards the *hilum* are syntropous.

If the radicle begins to decay, the seedling undergoes preemergence damping-off. This disease appears on the radicle as darkened spots. Eventually, it causes death of the seedling.

The plumule is the baby shoot. It grows after the radicle.

In 1880 Charles Darwin published a book about plants he had studied, The Power of Movement in Plants, where he mentions the radicle.

It is hardly an exaggeration to say that the tip of the radicle thus endowed [..] acts like the brain of one of the lower animals; the brain being situated within the anterior end of the body, receiving impressions from the sense-organs, and directing the several movements.

Hypocotyl

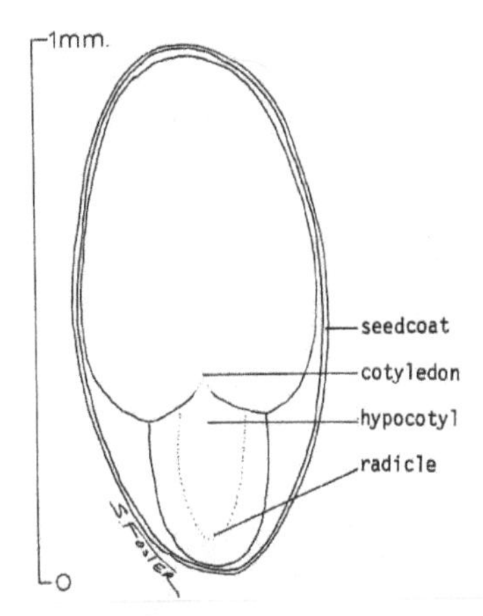

Diagram of Scouler's willow (*Salix scouleriana*) seed, indicating position of hypocotyl.

The hypocotyl (short for "hypocotyledonous stem", meaning "below seed leaf") is the stem of a germinating seedling, found below the cotyledons (seed leaves) and above the radicle (root).

Dicots

As the plant embryo grows at germination, it sends out a shoot called a radicle that becomes the primary root, and then penetrates down into the soil. After emergence of the radicle, the hypocotyl emerges and lifts the growing tip (usually including the seed coat) above the ground, bearing the embryonic leaves (called cotyledons), and the plumule that gives rise to the first true leaves. The hypocotyl is the primary organ of extension of the young plant and develops into the stem.

Monocots

The early development of a monocot seedling like cereals and other grasses is somewhat different. A structure called the coleoptile, essentially a part of the *cotyledon*, protects the young stem and plumule as growth pushes them up through the soil. A mesocotyl—that part of the young plant that lies between the seed (which remains buried) and the plumule—extends the shoot up to the soil surface, where secondary roots develop from just beneath the plumule. The primary root from the radicle may then fail to develop further. The mesocotyl is considered to be partly hypocotyl and partly cotyledon.

Not all monocots develop like the grasses. The onion develops in a manner similar to the first sequence described above, the seed coat and endosperm (stored food reserve) pulled upwards as the cotyledon extends. Later, the first true leaf grows from the node between the radicle and the sheath-like cotyledon, breaking through the cotyledon to grow past it.

Storage Organ

In some plants, the hypocotyl becomes enlarged as a storage organ. Examples include cyclamen, gloxinia and celeriac. In cyclamen this storage organ is called a tuber.

Hypocotyl Elongation Assay

One of the widely used assay in the field of photobiology is the investigation of the effect of changes in light quantity and quality on hypocotyl elongation. It is frequently used to study the growth promoting vs. growth repressing effects of application of plant hormones like ethylene. Under normal light conditions, hypocotyl growth is controlled by a process called photomorphogenesis, while shading the seedlings evokes a rapid transcriptional response which negatively regulates photomorphogenesis and results in increased rates of hypocotyl growth. This rate is highest when plants are kept in darkness mediated by a process called skotomorphogenesis, which contrasts photomorphogenesis.

Cotyledon

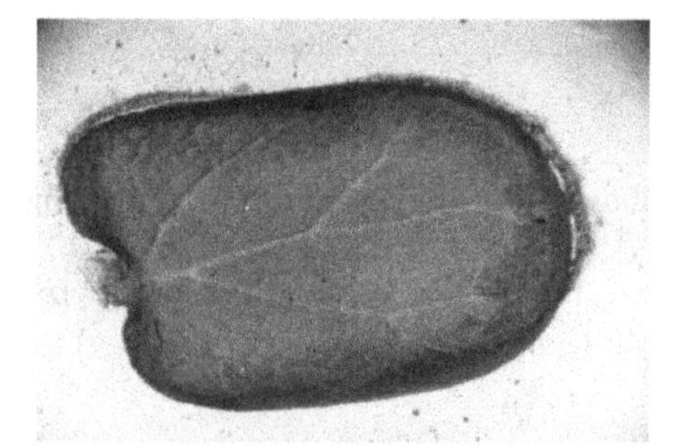

Cotyledon in formation before the accumulation of the reserve seen at Judas-tree (*Cercis siliquastrum*)

Comparison of a monocot and dicot sprouting. The visible part of the monocot plant (left) is actually the first true leaf produced from the meristem; the cotyledon itself remains within the seed

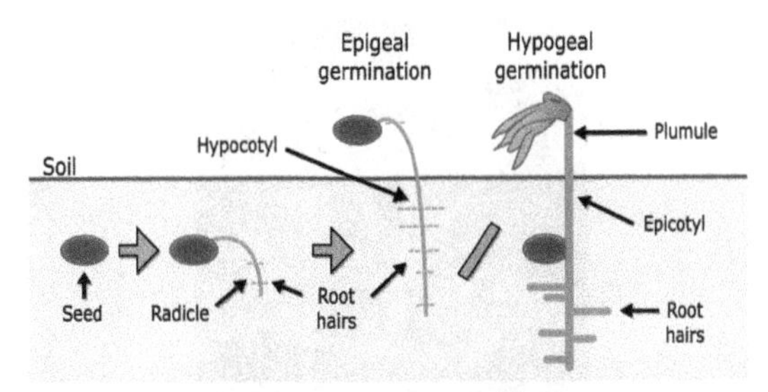

Schematic of epigeal vs hypogeal germination

A cotyledon is a significant part of the embryo within the seed of a plant, and is defined by the Oxford English Dictionary as "The primary leaf in the embryo of the higher plants (Phanerogams); the seed-leaf." Upon germination, the cotyledon may become the embryonic first leaves of a seedling. The number of cotyledons present is one characteristic used by botanists to classify the flowering plants (angiosperms). Species with one cotyledon are called monocotyledonous ("monocots"). Plants with two embryonic leaves are termed dicotyledonous ("dicots") and placed in the class Magnoliopsida.

Peanut seeds cut in half showing the embryos with cotyledons and primordial root.

In the case of dicot seedlings whose cotyledons are photosynthetic, the cotyledons are functionally similar to leaves. However, true leaves and cotyledons are developmentally distinct. Cotyledons are formed during embryogenesis, along with the root and shoot meristems, and are therefore present in the seed prior to germination. True leaves, however, are formed post-embryonically (i.e. after germination) from the shoot apical meristem, which is responsible for generating subsequent aerial portions of the plant.

Two-weeks-old cotyledons of Douglas-fir.

A seedling of Pinus halepensis with eight cotyledons

The cotyledon of grasses and many other monocotyledons is a highly modified leaf composed of a *scutellum* and a *coleoptile*. The scutellum is a tissue within the seed that is specialized to absorb stored food from the adjacent endosperm. The coleoptile is a protective cap that covers the *plumule* (precursor to the stem and leaves of the plant).

Gymnosperm seedlings also have cotyledons, and these are often variable in number (multicotyledonous), with from 2 to 24 cotyledons forming a whorl at the top of the hypocotyl (the embryonic stem) surrounding the plumule. Within each species, there is often still some variation in cotyledon numbers, e.g. Monterey pine (*Pinus radiata*) seedlings have 5–9, and Jeffrey pine (*Pinus jeffreyi*) 7–13 (Mirov 1967), but other species are more fixed, with e.g. Mediterranean cypress always having just two cotyledons. The highest number reported is for big-cone pinyon (*Pinus maximartinezii*), with 24 (Farjon & Styles 1997).

The cotyledons may be ephemeral, lasting only days after emergence, or persistent, enduring at least a year on the plant. The cotyledons contain (or in the case of gymnosperms and monocotyledons, have access to) the stored food reserves of the seed. As these reserves are used up, the cotyledons may turn green and begin photosynthesis, or may wither as the first true leaves take over food production for the seedling.

Epigeal Versus Hypogeal Development

Cotyledons may be either epigeal, expanding on the germination of the seed, throwing off the seed shell, rising above the ground, and perhaps becoming photosynthetic; or hypogeal, not expanding, remaining below ground and not becoming photosynthetic. The latter is typically the case where the cotyledons act as a storage organ, as in many nuts and acorns.

Hypogeal plants have (on average) significantly larger seeds than epigeal ones. They are also capable of surviving if the seedling is clipped off, as meristem buds remain underground (with epigeal plants, the meristem is clipped off if the seedling is grazed). The tradeoff is whether the plant should produce a large number of small seeds, or a smaller number of seeds which are more likely to survive.

Related plants show a mixture of hypogeal and epigeal development, even within the same plant family. Groups which contain both hypogeal and epigeal species include, for example, the Araucariaceae family of Southern Hemisphere conifers, the Fabaceae (pea family), and the genus *Lilium*. The frequently garden grown common bean - Phaseolus vulgaris - is epigeal while the closely related runner bean - Phaseolus coccineus - is hypogeal.

History

The term *cotyledon* was coined by Marcello Malpighi (1628–1694). John Ray was the first botanist to recognize that some plants have two and others only one, and eventually the first to recognize the immense importance of this fact to systematics, in *Methodus plantarum* (1682).

Theophrastus (3rd or 4th century BC) and Albertus Magnus (13th century) may also have recognized the distinction between the dicotyledons and monocotyledons.

Germination

Sunflower seedling, three days after germination

Germination rate testing on the germination table

Germination is the process by which a plant grows from a seed. The most common example of germination is the sprouting of a seedling from a seed of an angiosperm or gymnosperm. In addition, the growth of a sporeling from a spore, such as the spores of hyphae from fungal spores, is also germination. Thus, in a general sense, germination can be thought of as anything expanding into greater being from a small existence or germ.

Introduction

A seed tray used in horticulture for sowing and taking plant cuttings and growing plugs

Germination glass (glass sprouter jar) with a plastic sieve-lid

Brassica campestris germinating seeds

Germination is the growth of a plant contained within a seed; it results in the formation of the seedling, it is also the process of reactivation of metabolic machinery of the seed resulting in the emergence of radicle and plumule. The seed of a vascular plant is a small package produced in a fruit or cone after the union of male and female reproductive cells. All fully developed seeds contain an embryo and, in most plant species some store of food reserves, wrapped in a seed coat. Some plants produce varying numbers of seeds that lack embryos; these are called empty seeds and never germinate. Dormant seeds are ripe seeds that do not germinate because they are subject to external environmental conditions that prevent the initiation of metabolic processes and cell growth. Under proper conditions, the seed begins to germinate and the embryonic tissues resume growth, developing towards a seedling.

Seed germination depends on both internal and external conditions. The most important external factors include right temperature, water, oxygen or air and sometimes light or darkness. Various plants require different variables for successful seed germination. Often this depends on the individual seed variety and is closely linked to the ecological conditions of a plant's natural habitat. For some seeds, their future germination response is affected by environmental conditions during seed formation; most often these responses are types of seed dormancy.

- Water is required for germination. Mature seeds are often extremely dry and need to take in significant amounts of water, relative to the dry weight of the seed, before cellular metabolism and growth can resume. Most seeds need enough water to moisten the seeds but not enough to soak them. The uptake of water by seeds is called imbibition, which leads to the swelling and the breaking of the seed coat. When seeds are formed, most plants store a food reserve with the seed, such as starch, proteins, or oils. This food reserve provides nourishment to the growing embryo. When the seed imbibes water, hydrolytic enzymes are activated which break down these stored food resources into metabolically useful chemicals. After the seedling emerges from the seed coat and starts growing roots and leaves, the seedling's food reserves are typically exhausted; at this point photosynthesis provides the energy needed for continued growth and the seedling now requires a continuous supply of water, nutrients, and light.
- Oxygen is required by the germinating seed for metabolism. Oxygen is used in aerobic respiration, the main source of the seedling's energy until it grows leaves. Oxygen is an atmospheric gas that is found in soil pore spaces; if a seed is buried too deeply within the soil or the soil is waterlogged, the seed can be oxygen starved. Some seeds have impermeable seed coats that prevent oxygen from entering the seed, causing a type of physical dormancy which is broken when the seed coat is worn away enough to allow gas exchange and water uptake from the environment.
- Temperature affects cellular metabolic and growth rates. Seeds from different species and even seeds from the same plant germinate over a wide range of temperatures. Seeds often have a temperature range within which they will germinate, and they will not do so above or below this range. Many seeds germinate at temperatures slightly above 60-75 F (16-24 C) [room-temperature if you live in a centrally heated house], while others germinate just above freezing and others germinate only in response to alternations in temperature between warm and cool. Some seeds germinate when the soil is cool 28-40 F (-2 - 4 C), and some when the soil is warm 76-90 F (24-32 C). Some seeds require exposure to cold temperatures (vernalization) to break dormancy. Some seeds in a dormant state will not germinate even if conditions are favorable. Seeds that are dependent on temperature to end dormancy have a type of physiological dormancy. For example, seeds requiring the cold of winter are inhibited from germinating until they take in water in the fall and experience cooler temperatures. Four degrees Celsius is cool enough to end dormancy for most cool dormant seeds, but some groups, especially within the family Ranunculaceae and others, need conditions cooler than -5 C. Some seeds will only germinate after hot temperatures during a forest fire which cracks their seed coats; this is a type of physical dormancy.

Most common annual vegetables have optimal germination temperatures between 75-90 F (24-32 C), though many species (e.g. radishes or spinach) can germinate at significantly lower temperatures, as low as 40 F (4 C), thus allowing them to be grown from seeds in cooler climates. Suboptimal temperatures lead to lower success rates and longer germination periods.

- Light or darkness can be an environmental trigger for germination and is a type of physiological dormancy. Most seeds are not affected by light or darkness, but many seeds, including species found in forest settings, will not germinate until an opening in the canopy allows sufficient light for growth of the seedling.

Scarification mimics natural processes that weaken the seed coat before germination. In nature, some seeds require particular conditions to germinate, such as the heat of a fire (e.g., many Australian native plants), or soaking in a body of water for a long period of time. Others need to be passed through an animal's digestive tract to weaken the seed coat enough to allow the seedling to emerge.

Malted (germinated) barley grains

Dormancy

Some live seeds are dormant and need more time, and/or need to be subjected to specific environmental conditions before they will germinate. Seed dormancy can originate in different parts of the seed, for example, within the embryo; in other cases the seed coat is involved. Dormancy breaking often involves changes in membranes, initiated by dormancy-breaking signals. This generally occurs only within hydrated seeds. Factors affecting seed dormancy include the presence of certain plant hormones, notably abscisic acid, which inhibits germination, and gibberellin, which ends seed dormancy. In brewing, barley seeds are treated with gibberellin to ensure uniform seed germination for the production of barley malt.

Seedling Establishment

In some definitions, the appearance of the radicle marks the end of germination and the beginning of "establishment", a period that utilizes the food reserves stored in the seed. Germination and establishment as an independent organism are critical phases in the life of a plant when they are the most vulnerable to injury, disease, and water stress. The germination index can be used as an indicator of phytotoxicity in soils. The mortality between dispersal of seeds and completion of establishment can be so high that many species have adapted to produce huge numbers of seeds

Germination Rate and Germination Capacity

In agriculture and gardening, the germination rate describes how many seeds of a particular plant species, variety or seedlot are likely to germinate over a given period. It is a measure of germination time course and is usually expressed as a percentage, e.g., an 85% germination rate indicates that about 85 out of 100 seeds will probably germinate under proper conditions over the germination period given. The germination rate is useful for calculating the seed requirements for a given area or desired number of plants. In seed physiologists and seed scientists "germination rate" is the reciprocal of time taken for the process of germination to complete starting from time of sowing. On the other hand, the number of seed able to complete germination in a population (i.e. seed lot) is referred as germination capacity.

Germination of seedlings raised from seeds of eucalyptus after 3 days of sowing.

Dicot Germination

The part of the plant that first emerges from the seed is the embryonic root, termed the radicle or primary root. It allows the seedling to become anchored in the ground and start absorbing water. After the root absorbs water, an embryonic shoot emerges from the seed. This shoot comprises three main parts: the cotyledons (seed leaves), the section of shoot below the cotyledons (hypocotyl), and the section of shoot above the cotyledons (epicotyl). The way the shoot emerges differs among plant groups.

Epigeal

In epigeal germination (or epigeous germination), the *hypocotyl* elongates and forms a hook, pulling rather than pushing the cotyledons and apical meristem through the soil. Once it reaches the surface, it straightens and pulls the cotyledons and shoot tip of the growing seedlings into the air. Beans, tamarind and papaya are examples of plants that germinate this way.

Hypogeal

Germination can also be done by hypogeal germination (or hypogeous germination), where the epicotyl elongates and forms the hook. In this type of germination, the cotyledons stay underground where they eventually decompose. Peas, gram and mango, for example, germinate this way.

Monocot Germination

In monocot seeds, the embryo's radicle and cotyledon are covered by a coleorhiza and coleoptile, respectively. The coleorhiza is the first part to grow out of the seed, followed by the radicle. The coleoptile is then pushed up through the ground until it reaches the surface. There, it stops elongating and the first leaves emerge.

Precocious Germination

When a seed germinates without undergoing all four stages of seed development, i.e., globular, heart shape, torpedo shape, and cotyledonary stage, it is known as precocious germination.

Pollen Germination

Another germination event during the life cycle of gymnosperms and flowering plants is the germination of a pollen grain after pollination. Like seeds, pollen grains are severely dehydrated before being released to facilitate their dispersal from one plant to another. They consist of a protective coat containing several cells (up to 8 in gymnosperms, 2-3 in flowering plants). One of these cells is a tube cell. Once the pollen grain lands on the stigma of a receptive flower (or a female cone in gymnosperms), it takes up water and germinates. Pollen germination is facilitated by hydration on the stigma, as well as by the structure and physiology of the stigma and style. Pollen can also be induced to germinate *in vitro* (in a petri dish or test tube).

During germination, the tube cell elongates into a pollen tube. In the flower, the pollen tube then grows towards the ovule where it discharges the sperm produced in the pollen grain for fertilization. The germinated pollen grain with its two sperm cells is the mature male microgametophyte of these plants.

Self-incompatibility

Since most plants carry both male and female reproductive organs in their flowers, there is a high risk of self-pollination and thus inbreeding. Some plants use the control of pollen germination as a way to prevent this self-pollination. Germination and growth of the pollen tube involve molecular signaling between stigma and pollen. In self-incompatibility in plants, the stigma of certain plants can molecularly recognize pollen from the same plant and prevent it from germinating.

Spore Germination

Germination can also refer to the emergence of cells from resting spores and the growth of sporeling hyphae or thalli from spores in fungi, algae and some plants.

Conidia are asexual reproductive (reproduction without the fusing of gametes) spores of fungi which germinate under specific conditions. A variety of cells can be formed from the germinating conidia. The most common are germ tubes which grow and develop into hyphae. Another type of cell is a conidial anastomosis tube (CAT); these differ from germ tubes in that they are thinner, shorter, lack branches, exhibit determinate growth and home toward each other. Each cell is of a tubular shape, but the conidial anastomosis tube forms a bridge that allows fusion between conidia.

Resting Spores

In resting spores, germination that involves cracking the thick cell wall of the dormant spore. For example, in zygomycetes the thick-walled zygosporangium cracks open and the zygospore inside gives rise to the emerging sporangiophore. In slime molds, germination refers to the emergence of amoeboid cells from the hardened spore. After cracking the spore coat, further development involves cell division, but not necessarily the development of a multicellular organism (for example in the free-living amoebas of slime molds).

Ferns and Mosses

In plants such as bryophytes, ferns, and a few others, spores germinate into independent gametophytes. In the bryophytes (e.g., mosses and liverworts), spores germinate into protonemata, similar to fungal hyphae, from which the gametophyte grows. In ferns, the gametophytes are small, heart-shaped prothalli that can often be found underneath a spore-shedding adult plant.

Germination of Bacteria

Epigeal Germination

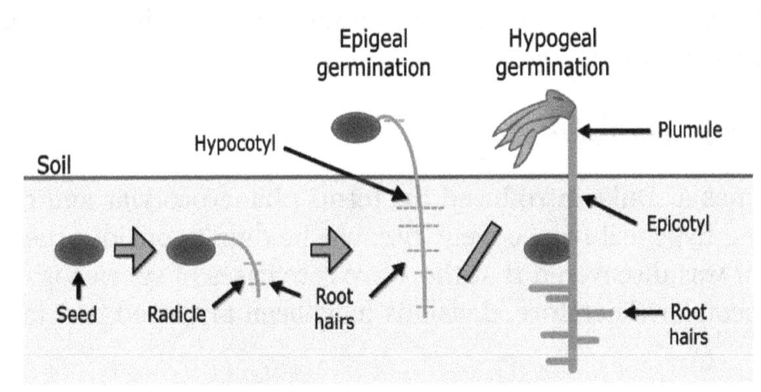

Epigeal vs. hypogeal germination

Epigeal germination is a botanical term indicating that the germination of a plant takes place above the ground. An example of a plant with epigeal germination is the common bean (*Phaseolus vulgaris*). The opposite of epigeal is hypogeal (underground germination).

Germination

Epigeal germination implies that the cotyledons are pushed above ground. The hypocotyl (part of the stem below the cotyledon) elongates while the epicotyl (part of the stem above the cotyledon) stays the same length. In this way, the hypocotyl pushes the cotyledon upward.

Normally, the cotyledon itself contains very little nutrients in plants that show this kind of germination. Instead, the first leaflets are already folded up inside it, and photosynthesis starts to take place in it rather quickly.

Because the cotyledon is positioned above the ground it is much more vulnerable to damage like night-frost or grazing. The evolutionary strategy is that the plant produces a large number of seeds, of which statistically a number survive.

Plants that show epigeal germination need external nutrients rather quickly in order to develop, so they are more frequent on nutrient-rich soils. The plants also need relatively much sunlight for photosynthesis to take place. Therefore they can be found more often in the field, at the border of forests, or as pioneer species.

Plants that show epigeal germination grow relatively fast, especially in the first phase when the leaflets unfold. Because of this, they occur frequently in areas that experience regular flooding, for example at the river borders in the Amazon region. The fast germination enables the plant to develop before the next flooding takes place. After the faster first phase, the plant develops more slowly than plants that show hypogeal germination.

It is possible that within the same genus one species shows epigeal germination while another species shows hypogeal germination. Some genera in which this happens are:

- *Phaseolus*: the common bean (*Phaseolus vulgaris*) shows epigeal germination, whereas the runner bean (*Phaseolus coccineus*) shows hypogeal germination
- *Lilium*
- *Araucaria*: species in the section *Eutacta* show epigeal germination, whereas species in the section *Araucaria* show hypogeal germination

Phanerocotylar vs. Cryptocotylar

In 1965, botanist James A. Duke introduced the terms phanerocotylar and cryptocotylar as synonyms for epigeal and hypogeal respectively, because he didn't consider these terms etymologically correct. Later, it was discovered that there are rare cases of species where the germination is epigeal and cryptocotylar. Therefore, divisions have been proposed that take both factors into account.

Hypogeal Germination

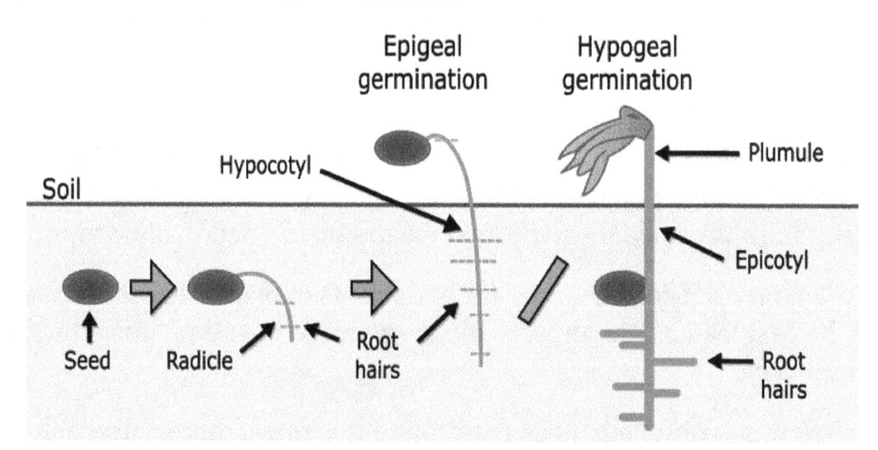

Epigeal vs. hypogeal germination

Hypogeal germination is a botanical term indicating that the germination of a plant takes place below the ground. An example of a plant with hypogeal germination is the pea (*Pisum sativum*). The opposite of hypogeal is epigeal (above-ground germination).

Germination

Hypogeal germination implies that the cotyledons stay below the ground. The epicotyl (part of the stem above the cotyledon) grows, while the hypocotyl (part of the stem below the cotyledon) stays the same length. In this way, the epicotyl pushes the plumule above the ground.

Normally, the cotyledon is fleshy, and contains many nutrients that are used for germination. No photosynthesis takes place within the cotyledon.

Because the cotyledon stays below the ground, it is much less vulnerable to for example night-frost or grazing. The evolutionary strategy is that the plant produces a relatively low number of seeds, but each seed has a bigger chance of surviving.

Plants that show hypogeal germination need relatively little in the way of external nutrients to grow, therefore they are more frequent on nutrient-poor soils. The plants also need less sunlight, so they can be found more often in the middle of forests, where there is much competition to reach the sunlight.

Plants that show hypogeal germination grow relatively slow, especially in the first phase. In areas that are regularly flooded, they need more time between floodings to develop. On the other hand, they are more resistant when a flooding takes place. After the slower first phase, the plant develops faster than plants that show epigeal germination.

It is possible that within the same genus one species shows hypogeal germination while another species shows epigeal germination. Some genera in which this happens are:

- *Phaseolus*: the runner bean (*Phaseolus coccineus*) shows hypogeal germination, whereas the common bean (*Phaseolus vulgaris*) shows epigeal germination
- *Lilium*
- *Araucaria*: species in the section *Araucaria* show hypogeal germination, whereas species in the section *Eutacta* show epigeal germination

Phanerocotylar vs. Cryptocotylar

In 1965, botanist James A. Duke introduced the terms cryptocotylar and phanerocotylar as synonyms for hypogeal and epigeal respectively, because he didn't consider these terms etymologically correct. Later, it was discovered that there are rare cases of species where the germination is epigeal and cryptocotylar. Therefore, divisions have been proposed that take both factors into account.

Lily Seed Germination Types

Lilies seed germination is classified as either epigeal or hypogeal. These classifications may be further refined as immediate or delayed. Whether a lily is epigeal or hypogeal may be related to survival strategies developed according to the climate where the lily originated. Epigeal lilies evolved in moderate climates. Hypogeal lilies evolved in harsher habitats where it would be advantageous to store food in a bulb, and later send up leaves in the spring.

Epigeal Lilies

Asiatic lilies include species of *Lilium lancifolium* (syn.tigrinum), *L. cernuum*, *L. davidii*, *L. maximowiczii*, *L. macultum*, *L. hollandicum*, *L. amabile*, *L. pumilum*, *L. concolor*, and *L. bulbiferum*. Epigeal trumpet lily species are *L. leucanthum*, *L. regale*, *L. sargentiae*, *L. sulphureum*, *L. rosthornii* and *L. henryi*. Many interdivisional hybrids also fall into the epigeal category. Epigeal lilies germinate under moist, warm conditions (approximately 70°F) in one stage, taking about 14 days. One stage means that they send up a leaf right away.

Hypogeal Lilies

Oriental lily species, such as *L. auratum*, *L. speciosum*, *L. nobilissimum*, *L. rubellum*, *L. alexandrae*, and *L. japonicum* and Martagon species *L. martagon*, *L. hansonii*, *L. medeoloides*, and *L. tsingtauense*, are all hypogeal. Hypogeal lilies require two or more stages with variations of temperature particular to each stage. For hypogeal lilies, the first stage of germination takes place entirely underground where the bulb is created. Hypogeals require a warm period of 3 months at 70°F, followed by a 3-month period at 40°F. A juvenile leaf appears in the second stage. The tiny bulbs are then planted in a warm area, usually outdoors.

Double Hypogeal Lilies

Double hypogeal lilies are the hardest to germinate and need multiple alternating periods of warm and cold. The exact sequence varies by species. Lilies that require these special conditions are often adapted to very specific conditions, and may be rare.

Sprouting

Mixed bean sprouts

Sprouting is the practice of germinating seeds to be eaten raw or cooked. Sprouts can be germinated at home or produced industrially. They are a prominent ingredient of the raw food diet and common in Eastern Asian cuisine.

Soybean sprouts on a plate

Sprouting, like cooking, reduces anti-nutritional compounds in raw legumes. Raw lentils, for example, contain lectins, antinutrional proteins which can be reduced by sprouting or cooking. Sprouting is also applied on a large scale to barley as a part of the malting process. A downside to consuming raw sprouts is that the process of germinating seeds can also be conducive to harmful bacterial growth.

Seeds Suitable for Sprouting

All viable seeds can be sprouted, but some sprouts should not be eaten raw. The most common food sprouts include:

- Pulses (legumes; pea family):

  alfalfa, clover, fenugreek, lentil, pea, chickpea, mung bean and soybean (bean sprouts).

- Cereals:

  oat, wheat, maize (corn), rice, barley, and rye

- Pseudocereals:

  quinoa, amaranth and buckwheat

- Oilseeds:

  sesame, sunflower, almond, hazelnut, hemp, linseed, peanut.

- Brassica (cabbage family)

  broccoli, cabbage, watercress, mustard, mizuna, radish, and daikon (*kaiware* sprouts), rocket (arugula), tatsoi, turnip.

- Umbelliferous vegetables (parsley family) - these may be used more as microgreens than sprouts.

  carrot, celery, fennel, parsley.

- Allium (onions) - cannot really distinguish between microgreens.

  onion, leek, green onion (*me-negi* in Japanese cuisine)

- Other vegetables and herbs:

  spinach, lettuce, milk thistle, lemon grass

Although whole oats can be sprouted, oat groats sold in food stores, which are dehulled and require steaming or roasting to prevent rancidity, will not sprout. Whole oats may have an indigestible hull which makes them difficult or even unfit for human consumption.

In the case of rice, the husk of the paddy is removed before sprouting. Brown rice is widely used for germination (GBR - Germinated Brown Rice) in Japan and other countries.

All the sprouts of the solanaceae (tomato, potato, paprika, aubergine or eggplant) and rhubarb cannot be eaten as sprouts, either cooked or raw, as they can be poisonous. Some sprouts can be cooked to remove the toxin, while others cannot.

With all seeds, care should be taken that they are intended for sprouting or human consumption rather than sowing. Seeds intended for sowing may be treated with chemical dressings. Several countries, such as New Zealand, also require that some varieties of imported edible seed be heat-treated, thus making them impossible to sprout. Quinoa in its natural state is very easy to sprout but when polished, or pre-cleaned of its saponin coating (becoming whiter), loses its power to germinate.

The Germination Process

Sprouting mung beans in a glass sprouter jar with a plastic sieve-lid

The germination process takes a few days and can be done at home manually, as a semi-automated process, or industrially on a large scale for commercial use.

Typically the seeds are first rinsed to remove soil and dirt and the mucilaginous substances produced by some seeds when they come in contact with water. Then they are soaked for 20 minutes to 12 hours, depending on the type and size of seed. The soaking increases the water content in the seeds and brings them out of quiescence. After draining and then rinsing seeds at regular intervals they germinate, or sprout.

For home sprouting, the seeds are soaked (big seeds) or moistened (small), then left at room temperature (13 to 21 °C or 55 to 70 °F) in a sprouting vessel. Many different types of vessels can be used. One type is a simple glass jar with a piece of cloth or nylon window screen secured over its rim. "Tiered" clear plastic sprouters are commercially available, allowing a number of "crops" to be grown simultaneously. By staggering sowings, a constant supply of young sprouts can be ensured. Any vessel used for sprouting must allow water to drain from it, because sprouts that sit in water will rot quickly. The seeds swell, may stick to the sides of the jar, and begin germinating within a day or two.

Another sprouting technique is to use a pulse drip method. The photo below on the right shows crimson clover sprouts grown on 1/8" urethane foam mats. It's a one-way watering system with micro sprinklers providing intermittent pulses of fresh water to reduce the risk of bacterial cross-contamination with Salmonella and E. coli during the sprouting process.

Crimson clover sprouts grown on 1/8" urethane foam mats with a pulse drip technique. Four micro-sprinklers cycle pulsing continuously over a 7-day period, each putting out about 1/2 gallon per hour. The four micro-sprinklers were each fitted with an LPD to keep the lines fully charged between pulses.

Sprouts are rinsed two to four times a day, depending on the climate and the type of seed, to provide them with moisture and prevent them from souring. Each seed has its own ideal sprouting time. After three to five days the sprouts will have grown to 5 to 8 centimetres (2–3 in) in length and will be suitable for consumption. If left longer they will begin to develop leaves, and are then known as baby greens. A popular baby green is sunflower after 7–10 days. Refrigeration can be used as needed to slow or halt the growth process of any sprout.

Common causes for sprouts to become inedible:

- Seeds are not rinsed well enough before soaking
- Seeds are left in standing water after the initial soaking
- Seeds are allowed to dry out
- Temperature is too high or too low
- Dirty equipment
- Insufficient air flow

- Contaminated water source
- Poor germination rate

Mung beans can be sprouted either in light or dark conditions. Those sprouted in the dark will be crisper in texture and whiter, as in the case of commercially available Chinese Bean Sprouts, but these have less nutritional content than those grown in partial sunlight. Growing in full sunlight is not recommended, because it can cause the beans to overheat or dry out. Subjecting the sprouts to pressure, for example, by placing a weight on top of them in their sprouting container, will result in larger, crunchier sprouts similar to those sold in grocery stores.

A very effective way to sprout beans like lentils or azuki is in colanders. Soak the beans in water for about 8 hours then place in the colander. Wash twice a day. The sprouted beans can be eaten raw or cooked.

Sprouting is also applied on a large scale to barley as a part of the malting process. Malted barley is an important ingredient in beer and is used in huge quantities. Most malted barley is distributed among wide retail sellers in North American regions.

Many varieties of nuts, such as almonds and peanuts, can also be started in their growth cycle by soaking and sprouting, although because the sprouts are generally still very small when eaten, they are usually called "soaks".

Nutritional Information

Sprouts used for a verrine.

Fresh mung bean sprouts in a bowl

Sprouts are said to be rich in digestible energy, bioavailable vitamins, minerals, amino acids, proteins, and phytochemicals, as these are necessary for a germinating plant to grow. These nutrients are essential for human health. The nutritional changes upon germination and sprouting are summarised below.

Chavan and Kadam (1989) concluded that

- "The desirable nutritional changes that occur during sprouting are mainly due to the breakdown of complex compounds into a more simple form, transformation into essential constituents and breakdown of nutritionally undesirable constituents."
- "The metabolic activity of resting seeds increases as soon as they are hydrated during soaking. Complex biochemical changes occur during hydration and subsequent sprouting. The reserve chemical constituents, such as protein, starch and lipids, are broken down by enzymes into simple compounds that are used to make new compounds."
- "Sprouting grains causes increased activities of hydrolytic enzymes, improvements in the contents of total proteins, fat, certain essential amino acids, total sugars, B-group vitamins, and a decrease in dry matter, starch and anti-nutrients. The increased contents of protein, fat, fibre and total ash are only apparent and attributable to the disappearance of starch. However, improvements in amino acid composition, B-group vitamins, sugars, protein and starch digestibilities, and decrease in phytates and protease inhibitors are the metabolic effects of the sprouting process."

Increases in Protein Quality Chavan and Kadam (1989) stated that "Very complex qualitative changes are reported to occur during soaking and sprouting of seeds. The conversion of storage proteins of cereal grains into albumins and globulins during sprouting may improve the quality of cereal proteins. Many studies have shown an increase in the content of the amino acid Lysine with sprouting."

"An increase in proteolytic activity during sprouting is desirable for nutritional improvement of cereals because it leads to hydrolysis of prolamins and the liberated amino acids such as glutamic and proline are converted to limiting amino acids such as lysine."

Increases in Crude Fibre content Cuddeford (1989), based on data obtained by Peer and Leeson (1985) stated that "In sprouted barley, crude fibre, a major constituent of cell walls, increases both in percentage and real terms, with the synthesis of structural carbohydrates, such as cellulose and hemicellulose". Chung et al. (1989) found that the fibre content increased from 3.75% in unsprouted barley seed to 6% in 5-day sprouts."

Crude Protein and Crude Fibre changes in Barley Sprouted over a 7-day period

	Crude Protein (% of DM)	Crude Fibre (% of DM)
Original seed	12.7%	5.4%
Day 1	12.7%	5.6%
Day 2	13.0%	5.9%

Day 3	13.6%	5.8%
Day 4	13.4%	7.4%
Day 5	13.9%	9.7%
Day 6	14.0%	10.8%
Day 7	15.5%	14.1%

Source: Cuddeford (1989), based on data obtained by Peer and Leeson (1985).

Increase of protein is not due to new protein being manufactured by the germination process but by the washing out of starch and conversion to fiber—increasing the relative proportion of protein.

Increases in Essential Fatty Acids

An increase in lipase activity has been reported in barley by MacLeod and White (1962), as cited by Chavan and Kadam (1989). Increased lipolytic activity during germination and sprouting causes hydrolysis of triacylglycerols to glycerol and constituent fatty acids.

Increases in Vitamin content According to Chavan and Kadam (1989), most reports agree that sprouting treatment of cereal grains generally improves their vitamin value, especially the B-group vitamins. Certain vitamins such as α-tocopherol (Vitamin-E) and β-carotene (Vitamin-A precursor) are produced during the growth process (Cuddeford, 1989).

According to Shipard (2005), "Sprouts provide a good supply of Vitamins A, E & C plus B complex. Like enzymes, vitamins serve as bioactive catalysts to assist in the digestion and metabolism of feeds and the release of energy. They are also essential for the healing and repair of cells. However, vitamins are very perishable, and in general, the fresher the feeds eaten, the higher the vitamin content. The vitamin content of some seeds can increase by up to 20 times their original value within several days of sprouting. Mung Bean sprouts have B vitamin increases, compared to the dry seeds, of - B_1 up 285%, B_2 up 515%, B_3 up 256%. Even soaking seeds overnight in water yields greatly increased amounts of B vitamins, as well as Vitamin C. Compared with mature plants, sprouts can yield vitamin contents 30 times higher."

Chelation of Minerals Shipard (2005) claims that - "When seeds are sprouted, minerals chelate or merge with protein, in a way that increases their function."

It is important to note that while these changes may sound impressive, the comparisons are between dormant non-sprouted seed to sprouted seed rather than comparisons of sprouts to mature vegetables. Compared to dry seeds there are very large increases in nutrients whereas compared with mature vegetables the increase is less. However, a sprout, just starting out in life, is likely to need and thus have more nutrients (percentage wise) than a mature vegetable.

Following table lists selected nutrients in kidney beans to show the effect of sprouting. Nutrients are calculated for 100 gms of non-water components to take the water out of equation since after sprouting beans absorb lot of water.

GABA Sprouting also have shown that can improve the levels of GABA, a compound involved in the regulation of blood pressure, and promoted the liberation of bioactive peptides in diverse legumes.

Nutrients in grams per 100 grams of non-water components of kidney beans				
Nutrient	**DRI**	**Raw Beans**	**Sprouted Beans**	**Ratio**
Protein (g)	50	26.72	45.15	1.7
Vitamin C (mg)	90	5.1	416	81.56
Thiamin (mg)	1.2	0.6	3.98	6.63
Riboflavin (mg)	1.3	.248	2.78	11.2
Niacin (mg)	16	2.33	31.40	13.47
Vitamin B-6 (mg)	1.3	0.45	0.91	2.02
Folate (μg)	400	446	634.25	1.422
Vitamin E(mg)	15	.25	0	0
Vitamin K (μg)	120	21.527	0	0

Health Concerns

Bacterial Infection

FDA health warning on a sprouts package

Commercially grown sprouts have been associated with multiple outbreaks of harmful bacteria, including salmonella and toxic forms of *Escherichia coli.* Such infections may be a result of contaminated seeds or of unhygienic production with high microbial counts. Sprout seeds can become contaminated in the fields where they are grown, and sanitizing steps may be unable to kill bacteria hidden in damaged seeds. A single surviving bacterium in a kilogram of seed can be enough to contaminate a whole batch of sprouts, according to the FDA.

To minimize the impact of the incidents and maintain public health, both the U.S. Food and Drug Administration (FDA) and Health Canada issued industry guidance on the safe manufacturing of edible sprouts and public education on their safe consumption. There are also publications for hobby farmers on safely growing and consuming sprouts at home. The recommendations include development and implementation of good agricultural practices and good manufacturing practices in the production and handling of seeds and sprouts, seed disinfection treatments, and microbial testing before the product enters the food supply.

In June 2011, contaminated fenugreek sprouts (grown from seed from Egypt) in Germany was identified as the source of the 2011 E. coli O104:H4 outbreak which the German officials had blamed wrongly, first on cucumbers from Spain and then on mung bean sprouts. In addition to Germany, where 3,785 cases and 45 deaths had been reported by the end of the outbreak, a handful of cases were reported in several countries including Switzerland, Poland, the Netherlands, Sweden, Denmark, the UK, Canada and the USA. Virtually all affected people had been in Germany shortly before becoming ill.

Antinutritional Factors

Some legumes, including sprouts, can contain toxins or antinutritional factors, which can be reduced by soaking, sprouting and cooking (e.g., stir frying). Joy Larkcom advises that to be on the safe side "one shouldn't eat large quantities of raw legume sprouts on a regular basis, no more than about 550g (20oz) daily".

Phytic acid, an antinutritional factor, occurs primarily in the seed coats and germ tissue of plant seeds. It forms insoluble or nearly insoluble compounds with many metal ions, including those of calcium, iron, magnesium and zinc, reducing their dietary availability. Diets high in phytic acid content and poor in these minerals produce mineral deficiency in experimental animals (Gontzea and Sutzescu, 1968, as cited in Chavan and Kadam, 1989). The latter authors state that the sprouting of cereals has been reported to decrease levels of phytic acid. Similarly, Shipard (2005) states that enzymes of germination and sprouting can help eliminate detrimental substances such as phytic acid. However, the amount of phytic acid reduction from soaking is only marginal, and not enough to counteract its antinutrient effects

Standards and Regulations

EU Regulation

In order to prevent incidents like the 2011 EHEC epidemic, the European Commission has issued three new, tightened regulations on March 11, 2013.

- Regulation (EU) No 208/2013*

The origin of the seeds has to be traceable always at all stages of processing, production and distribution. Therefore, a full description of the seeds or sprouts needs to be kept on record.

- Regulation (EU) No 209/2013*

This regulation amends Regulation (EC) No 2073/2005 in respect of microbiological criteria for sprouts and the sampling rules for poultry carcases and fresh poultry meat.

- Regulation (EU) No 211/2013*

Imported sprouts or seeds intended for the production of sprouts need a certificate according to the model declared in the Annex of this regulation. The certificate serves as proof that the production process complies with the general hygiene provisions in Part A of Annex I to Regulation (EC) No 852/2004 and the traceability requirements of Implementing Regulation (EU) No 208/2013.

Sprouted Bread

Vegan flourless sprouted wheat bread

Sprouted bread is a type of bread made from whole grains that have been allowed to sprout, that is, to germinate. There are a few different types of sprouted grain bread. Some are made with added flour, some are made with added gluten, and some, such as Essene bread, are made with very few additional ingredients.

Sprouted Breads

These are breads that contain the whole grain (or kernel, or berry) of various seeds after they have been sprouted. They are different from 'white' bread inasmuch as 'white' breads are made from ground wheat endosperm (after removal of the bran and germ). Whole grain breads include the bran, germ and endosperm, therefore providing more fiber, and naturally occurring vitamins and proteins. Sprouted (or germinated) grain breads have roughly the same amount of vitamins per gram.

A comparison of nutritional analyses shows that sprouted grains contain about 75% of the energy (carbohydrates), slightly higher protein and about 40% of the fat when compared to whole grains.

Wheat is not the only grain used in sprouted breads. Grains and legumes such as millet, barley, oat, lentil and soy may be used. Bread that is made from an array of grains and legumes can provide a complete set of amino acids, the building blocks of proteins. Sprouted breads may contain slightly more trace minerals and nutrients than non-sprouted breads. Other than that, they supply much the same advantages as whole grain breads over refined grain breads, such as lowered risk of coronary heart disease.

Sprouted wild-yeasted whole wheat bread

Essene Bread

Essene bread is a very primitive form of sprouted grain bread made from sprouted wheat and prepared at a low temperature. It is often eaten uncooked, or slightly heated, by proponents of raw foods. The Essenes, a Jewish religious group that flourished from the 2nd century BC to the 1st century AD, are credited with the technique and basic recipes for Essene bread, although no scholarly evidence exists for this claim. Sprouting and low-temperature preparation ensure the maximum possible vitamin content for this foodstuff. Sprouting also breaks down the lectins and other substances that some individuals may be sensitive or allergic to.

Essene bread - 70% sprouted rye, 30% spelt whole grain

Essene bread - 70% sprouted rye, 30% spelt whole grain

Essene flat bread - 100% sprouted wheat

Essene fruit bread - sprouted spelt

Seed Germinator

Gilbert White used hot beds warmed by manure to germinate melon seeds in England.

A seed germinator is a device for germinating seeds. Typically, these create an environment in which light, humidity and temperature are controlled to provide optimum conditions for the germination of seeds.

One type of germinator is the Copenhagen or Jacobsen tank. The seeds rest upon blotting paper which is kept moist by wicks which draw from a bath of water whose temperature is regulated. The humidity around each seed is kept high by means of glass funnels and a lid covering the tank.

Seed Dormancy

A dormant seed is one that is unable to germinate in a specified period of time under a combination of environmental factors that are normally suitable for the germination of the non-dormant seed. Dormancy is a mechanism to prevent germination during unsuitable ecological conditions, when the probability of seedling survival is low.

One important function of most seeds is delayed germination, which allows time for dispersal and prevents germination of all the seeds at the same time. The staggering of germination safeguards some seeds and seedlings from suffering damage or death from short periods of bad weather or from transient herbivores; it also allows some seeds to germinate when competition from other plants for light and water might be less intense. Another form of delayed seed germination is seed quiescence, which is different from true seed dormancy and occurs when a seed fails to germinate because the external environmental conditions are too dry or warm or cold for germination. Many species of plants have seeds that delay germination for many months or years, and some seeds can remain in the soil seed bank for more than 50 years before germination. Some seeds have a very long viability period, and the oldest documented germinating seed was nearly 2000 years old based on radiocarbon dating.

Overview

True dormancy or innate dormancy is caused by conditions within the seed that prevent germination under normally ideal conditions. Often seed dormancy is divided into two major categories based on what part of the seed produces dormancy: exogenous and endogenous. There are three types of dormancy based on their mode of action: physical, physiological and morphological.

There have been a number of classification schemes developed to group different dormant seeds, but none have gained universal usage. Dormancy occurs because of a wide range of reasons that often overlap, producing conditions in which definitive categorization is not clear. Compounding this problem is that the same seed that is dormant for one reason at a given point may be dormant for another reason at a later point. Some seeds fluctuate from periods of dormancy to non dormancy, and despite the fact that a dormant seed appears to be static or inert, in reality they are still receiving and responding to environmental cues.

Exogenous Dormancy

Exogenous dormancy is caused by conditions outside the embryo and is often broken down into three subgroups:

Physical Dormancy

Dormancy that is caused by an impermeable seed coat is known as physical dormancy. Physical dormancy is the result of impermeable layer(s) that develops during maturation and drying of the seed or fruit. This impermeable layer prevents the seed from taking up water or gases. As a result, the seed is prevented from germinating until dormancy is broken. In natural systems, physical dormancy is broken by several factors including high temperatures, fluctuating temperatures, fire, freezing/thawing, drying or passage through the digestive tracts of animals. Physical dormancy is believed to have developed >100 mya.

Once physical dormancy is broken it cannot be reinstated i.e. the seed is unable to enter secondary dormancy following unfavourable conditions unlike seeds with physiological dormancy mechanisms. Therefore, the timing of the mechanisms that breaks physical dormancy is critical and must be tuned to environmental cues. This maximises the chances for germination occurring in conditions where the plant will successfully germinate, establish and eventually reproduce.

Physical dormancy has been identified in the seeds of plants across 15 angiosperm families including:

- Anacardiaceae
- Bixaceae

- Cannaceae (monocot)
- Cistaceae
- Cochlospermaceae
- Convolvulaceae
- Cucurbitaceae
- Dipterocarpaceae
- Geraniaceae
- Legumeinosae
- Malvaceae
- Nelumbonaceae
- Rhamnaceae
- Sarcolaenaceae
- Sapindaceae

Physical dormancy has not been recorded in any gymnosperms

Generally, physical dormancy is the result of one or more palisade layers in the fruit or seed coat. These layers are lignified with malphigian cells tightly packed together and impregnated with water-repellent. In the Anacardiaceae and Nelumbonaceae families the seed coat is not well developed. Therefore, palisade layers in the fruit perform the functional role of preventing water uptake . While physical dormancy is a common feature, several species in these families do not have physical dormancy or produce non-dormant seeds.

Specialised structures, which function as a "water-gap", are associated with the impermeable layers of the seed to prevent the uptake of water. The water-gap is closed at seed maturity and is opened in response to the appropriate environmental signal. Breaking physical dormancy involves the disruption of these specialised structures within the seed, and acts as an environmental signal detector for germination. For example, legume (Fabaceae) seeds become permeable after the thin-walled cells of lens (water-gap structure). Following disrupted pulls apart to allow water entry into the seed.Other water gap structures include carpellary micropyle, bixoid chalazal plug, imbibition lid and the suberised 'stopper'.

In nature, the seed coats of physically dormant seeds are thought to become water permeable over time through repeated heating and cooling over many months-years in the soil seedbank. For example, the high and fluctuating temperatures during the dry season in northern Australia promote dormancy break in impermeable seeds of *Stylosanthes humilis* and *S.hamata* (Fabaceae).

Mechanical Dormancy

Mechanical dormancy occurs when seed coats or other coverings are too hard to allow the embryo to expand during germination. In the past this mechanism of dormancy was ascribed to a number of species that have been found to have endogenous factors for their dormancy instead. These endogenous factors include low embryo growth potential.

Chemical Dormancy

Includes growth regulators etc., that are present in the coverings around the embryo. They may be leached out of the tissues by washing or soaking the seed, or deactivated by other means. Other chemicals that prevent germination are washed out of the seeds by rainwater or snow melt.

Endogenous Dormancy

Endogenous dormancy is caused by conditions within the embryo itself, and it is also often broken down into three subgroups: physiological dormancy, morphological dormancy and combined dormancy, each of these groups may also have subgroups.

Physiological Dormancy

Physiological dormancy prevents embryo growth and seed germination until chemical changes occur. These chemicals include inhibitors that often retard embryo growth to the point where it is not strong enough to break through the seed coat or other tissues. Physiological dormancy is indicated when an increase in germination rate occurs after an application of gibberellic acid (GA3) or after Dry after-ripening or dry storage. It is also indicated when dormant seed embryos are excised and produce healthy seedlings: or when up to 3 months of cold (0–10 °C) or warm (=15 °C) stratification increases germination: or when dry after-ripening shortens the cold stratification period required. In some seeds physiological dormancy is indicated when scarification increases germination.

Physiological dormancy is broken when inhibiting chemicals are broken down or are no longer produced by the seed; often by a period of cool moist conditions, normally below (+4C) 39F, or in the case of many species in *Ranunculaceae* and a few others,(−5C) 24F. Abscisic acid is usually the growth inhibitor in seeds and its production can be affected by light. Some plants like Peony species have multiple types of physiological dormancy, one affects radicle (root) growth while the other affects plumule (shoot) growth. Seeds with physiological dormancy most often do not germinate even after the seed coat or other structures that interfere with embryo growth are removed. Conditions that affect physiological dormancy of seeds include:

- Drying; some plants including a number of grasses and those from seasonally arid regions need a period of drying before they will germinate, the seeds are released but need to have a lower moisture content before germination can begin. If the seeds remain moist after dispersal, germination can be delayed for many months or even years. Many herbaceous plants from temperate climate zones have physiological dormancy that disappears with drying of the seeds. Other species will germinate after dispersal only under very narrow temperature ranges, but as the seeds dry they are able to germinate over a wider temperature range.
- Photodormancy or light sensitivity affects germination of some seeds. These photoblastic seeds need a period of darkness or light to germinate. In species with thin seed coats, light

may be able to penetrate into the dormant embryo. The presence of light or the absence of light may trigger the germination process, inhibiting germination in some seeds buried too deeply or in others not buried in the soil.

- Thermodormancy is seed sensitivity to heat or cold. Some seeds including cocklebur and amaranth germinate only at high temperatures (30C or 86F). Many plants that have seeds that germinate in early to mid summer have thermodormancy and germinate only when the soil temperature is warm. Other seeds need cool soils to germinate, while others like celery are inhibited when soil temperatures are too warm. Often thermodormancy requirements disappear as the seed ages or dries.

Seeds are classified as having deep physiological dormancy under these conditions: applications of GA3 does not increase germination; or when excised embryos produce abnormal seedlings; or when seeds require more than 3 months of cold stratification to germinate.

Morphological Dormancy

In morphological dormancy, the embryo is underdeveloped or undifferentiated. Some seeds have fully differentiated embryos that need to grow more before seed germination, or the embryos are not differentiated into different tissues at the time of fruit ripening.

- Immature embryos – some plants release their seeds before the tissues of the embryos have fully differentiated, and the seeds ripen after they take in water while on the ground, germination can be delayed from a few weeks to a few months.

Combined Dormancy

Seeds have both morphological and physiological dormancy.

- Morpho-physiological or morphophysiological dormancy occurs when seeds with underdeveloped embryos, also have physiological components to dormancy. These seeds therefore require dormancy-breaking treatments as well as a period of time to develop fully grown embryos.
- Intermediate simple
- Deep simple
- Deep simple epicotyl
- Deep simple double
- Intermediate complex
- Deep complex

Combinational Dormancy

Combinational dormancy occurs in some seeds, where dormancy is caused by both exogenous (physical) and endogenous (physiological) conditions. some *Iris* species have both hard impermeable seeds coats and physiological dormancy.

Secondary Dormancy

Secondary dormancy occurs in some non-dormant and post dormant seeds that are exposed to conditions that are not favorable for germination, like high temperatures. It is caused by conditions that occur after the seed has been dispersed. The mechanisms of secondary dormancy are not yet fully understood but might involve the loss of sensitivity in receptors in the plasma membrane.

Not all seeds undergo a period of dormancy, many species of plants release their seeds late in the year when the soil temperature is too low for germination or when the environment is dry. If these seeds are collected and sown in an environment that is warm enough, and/or moist enough, they will germinate. Under natural conditions non dormant seeds released late in the growing season wait until spring when the soil temperature rises or in the case of seeds dispersed during dry periods until it rains and there is enough soil moisture.

Seeds that do not germinate because they have fleshy fruits that retard germination are quiescent, not dormant.

Many garden plants have seeds that will germinate readily as soon as they have water and are warm enough, though their wild ancestors had dormancy. These cultivated plants lack seed dormancy because of generations of selective pressure by plant breeders and gardeners that grew and kept plants that lacked dormancy.

Seeds of some mangroves are viviparous and begin to germinate while still attached to the parent; they produce a large, heavy root, which allows the seed to penetrate into the ground when it falls.

Seed Dispersal

Seed dispersal is the movement or transport of seeds away from the parent plant. Plants have very limited mobility and consequently rely upon a variety of dispersal vectors to transport their propagules, including both abiotic and biotic vectors. Seeds can be dispersed away from the parent plant individually or collectively, as well as dispersed in both space and time. The patterns of seed dispersal are determined in large part by the dispersal mechanism and this has important implications for the demographic and genetic structure of plant populations, as well as migration patterns and species interactions. There are five main modes of seed dispersal: gravity, wind, ballistic, water, and by animals. Some plants are serotinous and only disperse their seeds in response to an environmental stimulus.

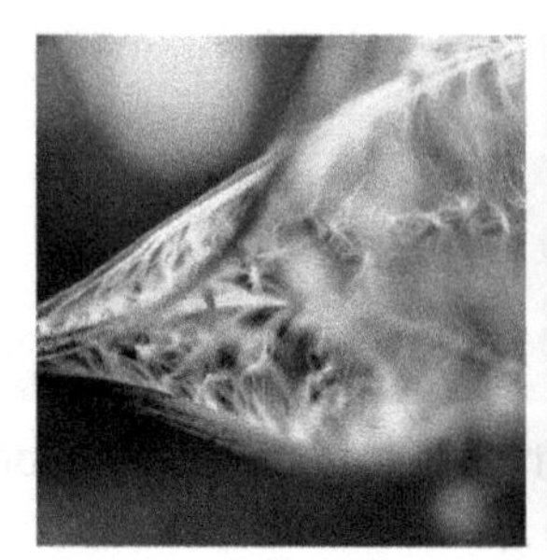
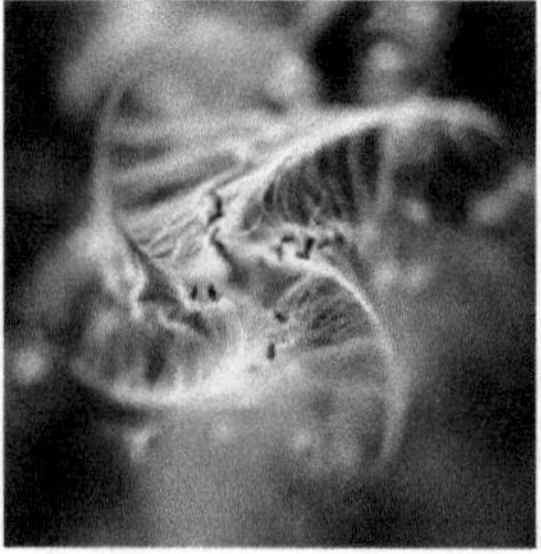

Epilobium hirsutum seed head dispersing seeds

Benefits

Seed dispersal is likely to have several benefits for plant species. First, seed survival is often higher away from the parent plant. This higher survival may result from the actions of density-dependent seed and seedling predators and pathogens, which often target the high concentrations of seeds beneath adults. Competition with adult plants may also be lower when seeds are transported away from their parent.

Seed dispersal also allows plants to reach specific habitats that are favorable for survival, a hypothesis known as directed dispersal. For example, *Ocotea endresiana* (Lauraceae) is a tree species from Latin America which is dispersed by several species of birds, including the three-wattled bellbird. Male bellbirds perch on dead trees in order to attract mates, and often defecate seeds beneath these perches where the seeds have a high chance of survival because of high light conditions and escape from fungal pathogens. In the case of fleshy-fruited plants, seed-dispersal in animal guts (endozoochory) often enhances the amount, the speed, and the asynchrony of germination, which can have important plant benefits.

Seeds dispersed by ants (myrmecochory) are not only dispersed short distances but are also buried underground by the ants. These seeds can thus avoid adverse environmental effects such as fire or drought, reach nutrient-rich microsites and survive longer than other seeds. These features are peculiar to myrmecochory, which may thus provide additional benefits not present in other dispersal modes.

Finally, at another scale, seed dispersal may allow plants to colonize vacant habitats and even new geographic regions.

Types

Seed dispersal is sometimes split into *autochory* (when dispersal is attained using the plant's own means) and *allochory* (when obtained through external means).

Autochory

Autochorous plants disperse their seed without any help from an external vector, as a result this limits plants considerably as to the distance they can disperse their seed. Two other types of autochory not described in detail here are blastochory, where the stem of the plant crawls along the ground to deposit its seed far from the base of the plant, and herpochory (the seed crawls by means of trichomes and changes in humidity).

Gravity

Barochory or the plant use of gravity for dispersal is a simple means of achieving seed dispersal. The effect of gravity on heavier fruits causes them to fall from the plant when ripe. Fruits exhibiting this type of dispersal include apples, coconuts and passionfruit and those with harder shells (which often roll away from the plant to gain more distance). Gravity dispersal also allows for later transmission by water or animal.

Ballistic Dispersal

Ballochory is a type of dispersal where the seed is forcefully ejected by explosive dehiscence of the fruit. Often the force that generates the explosion results from turgor pressure within the fruit or due to internal tensions within the fruit. Some examples of plants which disperse their seeds autochorously include: *Impatients spp., Arceuthobium spp., Ecballium spp., Geranium spp., Cardamine hirsuta* and others. An exceptional example of ballochory is *Hura crepitans*–this plant is commonly called the dynamite tree due to the sound of the fruit exploding. The explosions are powerful enough to throw the seed up to 100 meters.

Allochory

Allochory refers to any of many types of seed dispersal where a vector or secondary agent is used to disperse seeds. This vectors may include wind, water, animals or others.

Wind

Wind dispersal of dandelion seeds

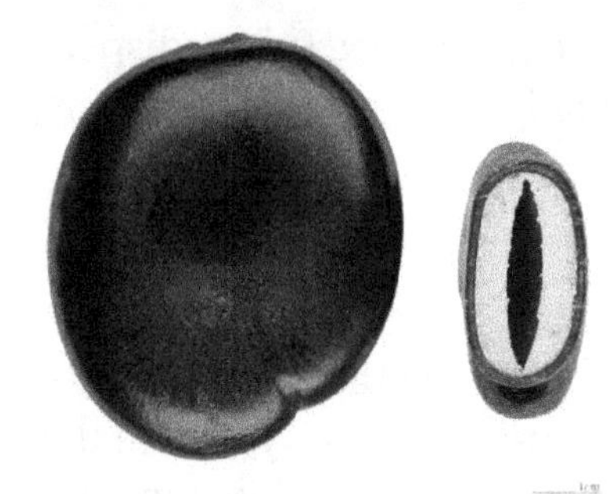

Entada phaseoloides – Hydrochory

Wind dispersal (*anemochory*) is one of the more primitive means of dispersal. Wind dispersal can take on one of two primary forms: seeds can float on the breeze or alternatively, they can flutter to the ground. The classic examples of these dispersal mechanisms, in the temperate northern hemisphere, include dandelions, which have a feathery pappus attached to their seeds and can be dispersed long distances, and maples, which have winged seeds (samaras) and flutter to the ground. An important constraint on wind dispersal is the need for abundant seed production to maximize the likelihood of a seed landing in a site suitable for germination. There are also strong evolutionary constraints on this dispersal mechanism. For instance, Cody and Overton (1996) found that species in the Asteraceae on islands tended to have reduced dispersal capabilities (i.e., larger seed mass and smaller pappus) relative to the same species on the mainland. Also, *Helonias bullata*, a species perennial herb native to the United States, evolved to utilize wind dispersal as the primary seed dispersal mechanism; however, limited wind in its habitat prevents the seeds to successfully disperse away from its parents, resulting in clusters of population. Reliance on wind dispersal is common among many weedy or ruderal species. Unusual mechanisms of wind dispersal include tumbleweeds, where the entire plant is blown by the wind. Physalis fruits, when not fully ripe, may sometimes be dispersed by wind due to the space between the fruit and the covering calyx which acts as air bladder.

Water

Many aquatic (water dwelling) and some terrestrial (land dwelling) species use *hydrochory*, or seed dispersal through water. Seeds can travel for extremely long distances, depending on the specific mode of water dispersal; this especially applies to fruits which are waterproof and float.

The water lily is an example of such a plant. Water lilies' flowers make a fruit that floats in the water for a while and then drops down to the bottom to take root on the floor of the pond. The seeds of palm trees can also be dispersed by water. If they grow near oceans, the seeds can be transported by ocean currents over long distances, allowing the seeds to be dispersed as far as other continents.

Mangrove trees grow directly out of the water; when their seeds are ripe they fall from the tree and grow roots as soon as they touch any kind of soil. During low tide, they might fall in soil instead of water and start growing right where they fell. If the water level is high, however, they can be carried far away from where they fell. Mangrove trees often make little islands as dirt and other things collect in their roots, making little bodies of land.

A special review for oceanic waters hydrochory can be seen at oceanic dispersal.

The "bill" and seed dispersal mechanism of *Geranium pratense*

By Animals

Animals can disperse plant seeds in several ways, all named *zoochory*. Seeds can be transported on the outside of vertebrate animals (mostly mammals), a process known as *epizoochory*. Plant species transported externally by animals can have a variety of adaptations for dispersal, including adhesive mucus, and a variety of hooks, spines and barbs. A typical example of an epizoochorous plant is *Trifolium angustifolium*, a species of Old World clover which adheres to animal fur by means of stiff hairs covering the seed. Epizoochorous plants tend to be herbaceous plants, with many representative species in the families Apiaceae and Asteraceae. However, epizoochory is a relatively rare dispersal syndrome for plants as a whole; the percentage of plant species with seeds adapted for transport on the outside of animals is estimated to be below 5%. Nevertheless, epizoochorous transport can be highly effective if seeds attach to wide-ranging animals. This form of seed dispersal has been implicated in rapid plant migration and the spread of invasive species.

The small hooks on the surface of a bur enable attachment to animal fur for dispersion.

Seed dispersal via ingestion by vertebrate animals (mostly birds and mammals), or *endozoochory*, is the dispersal mechanism for most tree species. Endozoochory is generally a coevolved mutualistic relationship in which a plant surrounds seeds with an edible, nutritious fruit as a good food for animals that consume it. Birds and mammals are the most important seed dispersers, but a wide variety of other animals, including turtles and fish, can transport viable seeds. The exact percentage of tree species dispersed by endozoochory varies between habitats, but can range to over 90% in some tropical rainforests. Seed dispersal by animals in tropical rainforests has received much attention, and this interaction is considered an important force shaping the ecology and evolution of vertebrate and tree populations. In the tropics, large animal seed dispersers (such as tapirs, chimpanzees and hornbills) may disperse large seeds with few other seed dispersal agents. The extinction of these large frugivores from poaching and habitat loss may have negative effects on the tree populations that depend on them for seed dispersal.

Seed dispersal by ants (*myrmecochory*) is a dispersal mechanism of many shrubs of the southern hemisphere or understorey herbs of the northern hemisphere. Seeds of myrmecochorous plants have a lipid-rich attachment called the elaiosome, which attracts ants. Ants carry such seeds into their colonies, feed the elaiosome to their larvae and discard the otherwise intact seed in an underground chamber. Myrmecochory is thus a coevolved mutualistic relationship between plants and seed-disperser ants. Myrmecochory has independently evolved at least 100 times in flowering plants and is estimated to be present in at least 11 000 species, but likely up to 23 000 or 9% of all species of flowering plants. Myrmecochorous plants are most frequent in the fynbos vegetation of the Cape Floristic Region of South Africa, the kwongan vegetation and other dry habitat types of Australia, dry forests and grasslands of the Mediterranean region and northern temperate forests of western Eurasia and eastern North America, where up to 30–40% of understorey herbs are myrmecochorous.

Seed predators, which include many rodents (such as squirrels) and some birds (such as jays) may also disperse seeds by hoarding the seeds in hidden caches. The seeds in caches are usually well-protected from other seed predators and if left uneaten will grow into new plants. In addition, rodents may also disperse seeds via seed spitting due to the presence of secondary metabolites in ripe fruits. Finally, seeds may be secondarily dispersed from seeds deposited by primary animal dispersers. For example, dung beetles are known to disperse seeds from clumps of feces in the process of collecting dung to feed their larvae.

Other types of zoochory are *chiropterochory* (by bats), *malacochory* (by molluscs, mainly terrestrial snails), *ornithochory* (by birds) and *saurochory* (by non-bird sauropsids).

By Humans

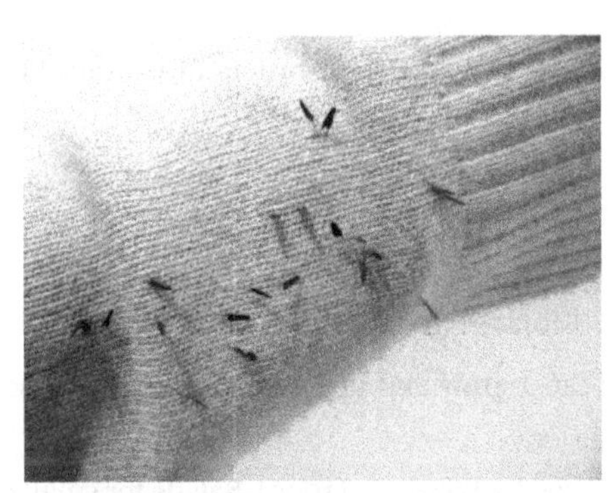

Epizoochory in *Bidens tripartita*; the seeds have attached to the clothes of a human.

Dispersal by humans (*anthropochory*) used to be seen as a form of dispersal by animals. Recent research points out that human dispersers differ from animal dispersers by a much higher mobility based on the technical means of human transport. Dispersal by humans on the one hand may act on large geographical scales and lead to invasive species. On the other hand, dispersal by humans also acts on smaller, regional scales and drives the dynamics of existing biological populations. Humans may disperse seeds by many various means and some surprisingly high distances have been repeatedly measured. Examples are: dispersal on human clothes (up to 250 m), on shoes (up to 5 km) or by cars (regularly ~ 250 m, singles cases > 100 km).

Deliberate seed dispersal also occurs as seed bombing. This has risks as unsuitable provenance may introduce genetically unsuitable plants to new environments.

Consequences

Seed dispersal has many consequences for the ecology and evolution of plants. Dispersal is necessary for species migrations, and in recent times dispersal ability is an important factor in whether or not a species transported to a new habitat by humans will become an invasive species. Dispersal is also predicted to play a major role in the origin and maintenance of species diversity. For example, myrmecochory increased the rate of diversification more than twofold in plant groups in which it has evolved because myrmecochorous lineages contain more than twice as many species as their non-myrmecochorous sister groups. Dispersal of seeds away from the parent organism has a central role in two major theories for how biodiversity is maintained in natural ecosystems, the Janzen-Connell hypothesis and recruitment limitation. Seed dispersal is essential in allowing forest migration of flowering plants.

In addition, the speed and direction of wind are highly influential in the dispersal process and in turn the deposition patterns of floating seeds in the stagnant water bodies. The transportation of seeds is led by the wind direction. This effects colonization situated on the banks of a river or to wetlands adjacent to streams relative to the distinct wind directions. The wind dispersal process can also effect connections between water bodies. Essentially, wind plays a larger role in the dispersal of waterborne seeds in a short period of time, days and seasons, but the ecological process allows the process to become balanced throughout a time period of several years. The time period of which the dispersal occurs is essential when considering the consequences of wind on the ecological process.

References

- Buczacki, S. and Harris, K., Pests, Diseases & Disorders of Garden Plants, HarperCollins, 1998, p116 ISBN 0-00-220063-5
- Raven, Peter H.; Ray F. Evert; Susan E. Eichhorn (2005). Biology of Plants, 7th Edition. New York: W.H. Freeman and Company Publishers. pp. 504–508. ISBN 0-7167-1007-2.
- Derek Bewley, J.; Black, Michael; Halmer, Peter (2006). The encyclopedia of seeds: science, technology and uses Cabi Series. CABI. p. 203. ISBN 0-85199-723-6. Retrieved 2009-08-28.
- Shipard, Isabell (2005). How can I grow and use sprouts as living food?. [Nambour, Qld.?]: David Stewart. ISBN 9780975825204.
- Douglass, Joy Larkcom ; illustrated by Elizabeth (1995). Salads for small gardens (2nd ed.). [London]: Hamlyn. ISBN 0-600-58509-3.
- Natural Antinutritive Substances in Foodstuffs and Forages (1 ed.). S. Karger; 1 edition (August 28, 1968). p. 184. ISBN 978-3805508568.
- Shipard, Isabell (2005). How can I grow and use sprouts as living food?. [Nambour, Qld.?]: David Stewart. ISBN 0975825208.
- Richard Mabey (2007), Gilbert White: a biography of the author of The natural history of Selborne, pp. 58–61, ISBN 978-0-8139-2649-0
- Fenner, Michael; Thompson, Ken (2005). The ecology of seeds. Publisher Cambridge University Press. p. 98. ISBN 978-0-521-65368-8. Retrieved 2009-08-15.
- Fenner, Michael; Thompson, Ken (2005), "The ecology of seeds", Publisher Cambridge University Press: 97, ISBN 978-0-521-65368-8, retrieved 2009-08-15
- Baskin, Carol C. Baskin, Jerry M. (2014). Seeds : ecology, biogeography, and evolution of dormancy and germination (Second edition. ed.). Amsterdam: Elsevier. ISBN 978-0-12-416677-6.

3

Essential Elements of Seed Science

Seed contamination is the process in which seeds are mixed for the purpose of agriculture. For example, mixing corn seed with the seed of weed. The alternative elements of seed science are seed production and gene diversity, seed enhancement, seed paper, seed testing, speed limit enforcement, seed drill etc. The topics discussed in the chapter are of great importance to broaden the existing knowledge on seed science.

Seed Production and Gene Diversity

Genetic diversity is often a major consideration in e g forest crops.

Group Coancestry of a Population

Consider the gene pool of a seed orchard crop or other source of seeds with parents. The gene pool is large as there are many seeds in a seed crop, so there is no genetic drift. The probability that the first gene originates from genotype i in the seed orchard is p_i, and the probability that the second originates from genotype j is p_j. The probability that these two genes originate from the orchard genotypes i and j are identical by descent (IBD) is θ_{ij}. This is the Malecot's method of coancestry (or "coefficient of kinship"; "coefficient of relationship" is a similar measure which can be computed) between genotype i and j. The probability that any pair of genes is IBD, Θ, can be found by adding over all possible pairs of genes from N parents. Formula for the group coancestry (from which gene diversity can be obtained) of seed orchard crops.

$$\Theta=\sum_{i=1}^{N}\sum_{j=1}^{N} p_i p_j \theta_{ij}.$$

The group coancestry of a seed orchard crop can be divided in two terms, one for self-coancestry and one for cross-coancestry.

$$\Theta=\sum_{i=1}^{N} p_i^2(1+F_i)/2 + \sum_{i=1}^{N}\sum_{j=1}^{N} p_i p_j \theta_{ij},$$ the last summation excluding j=i.

Let's consider a simple case. If the seed orchard genotypes are unrelated the second term is zero, if there is no inbreeding, the first term becomes simple. For no relatedness and no inbreeding, status number (N_s, effective number of parents) becomes

$$N_S = 1/\sum_{i=1}^{N} p_i^2$$

Similar but less developed expressions has been used many times before, it has similarities to the concept of effective population size as defined by.

Seed Contamination

Foreign seeds can be caught anywhere, including these harvested flax seeds

Seed contamination is the mixing of seeds used for agriculture with other seeds which are not desirable or soil (which may carry seeds). An example would be mixing corn seed with weed seed. These contaminant seeds can be either common weeds or other crop seeds.

In genetic engineering, seed contamination may refer to unwanted genes found in a seed, or pieces of DNA from other genetically engineered seeds.

Contamination in Genetic Engineering

Contamination in this area is most likely caused by other genetically engineered seeds cross pollinating with unchanged seeds. The new seeds that are produced will have significantly altered DNA than the parent seed. This mutation could potentially then be passed down for generations. It is possible that the alterations could affect plant growth and development.

Prevention

Preventing seed contamination in agriculture is fairly difficult, almost impossible. Better machines constantly help to reduce the number of foreign seeds gathered.

Machines, such as this tractor, help gather seeds. As they improve, the amount of foreign seeds they may pick up decreases.

Preventing seed contamination in genetic engineering is simply a matter of separating "pure" seeds from contaminated ones. This will keep a supply of unaltered seeds for future use. If they are allowed to cross pollinate, it is possible that the altered traits will be passed down.

Government agencies, especially within the United States, are setting up programs to stop seed contamination from genetic engineering. The *Gone to Seed* report lays out a plan to:

1. Conduct a study to determine the effects, causes, and extent of seed contamination
2. Change current seed regulations to prevent contamination
3. Create a "bank" of unaltered seeds

Seed Enhancement

Seed enhancement is a range of treatments of seeds that improves their performance after harvesting and conditioned, but before they are sown. They include priming, steeping, hardening, pregermination, pelleting, encrusting, film-coating, tagging and others, but excludes treatments for control of seed born pathogens.

They are used to improve seed sowing, germination and seedling growth by altering seed vigor and/or the physiological state of the seed. The alteration may improve vigor or the physiological state of the seed by enhancing uniformity of germination.

Treatments may include hydration treatments, such as priming, steeping, hardening, and pre-germination. Other treatment include the use of chemicals that trigger systemic acquired resistance or improve stress tolerance. The use of antioxidants. Enhancements like pelleting, coating and encrusting improve seed handling and planting. Some treatments enhance nutrient availability or provide inoculates by delivering materials (other than pesticides) needed during sowing, germination and seedling establishment.

Seed Saving

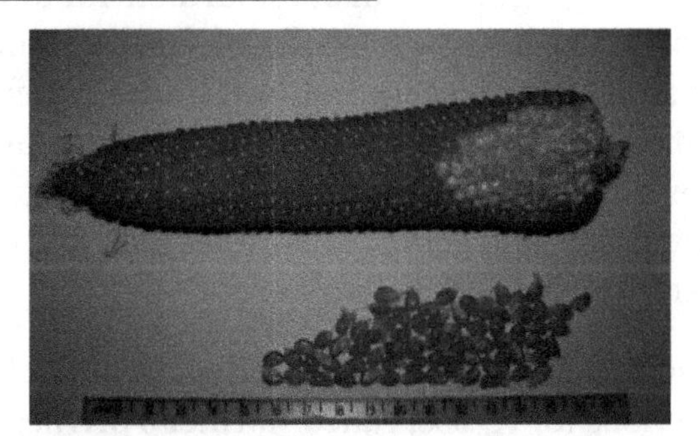

Partially shelled popcorn seed saved for planting

In agriculture and gardening, seed saving (sometimes known as brown bagging) is the practice of saving seeds or other reproductive material (e.g. tubers) from vegetables, grain, herbs, and flowers for use from year to year for annuals and nuts, tree fruits, and berries for perennials and trees. This is the traditional way farms and gardens were maintained for the last 12,000 years.

In recent decades, beginning in the latter part of the 20th century, there has been a major shift to purchasing seed annually from commercial seed suppliers. Much of the grassroots seed-saving activity today is the work of home gardeners.

Method

To be successful at seed saving, new skills need to be developed to ensure that desired characteristics are retained in the landraces of the plant variety. Important considerations are the separation distance needed from plants of the same species to ensure that cross-pollination with another variety does not occur, and the minimum number of plants to be grown which will preserve inherent genetic diversity. It is also necessary to recognize the preferred characteristics of the cultivar being grown so that plants that are not breeding true are selected against, and to understand the breeding of improvements to the cultivar. Diseases that are seed-borne must be recognized so that they can be eliminated. Seed storage methods must be good enough to maintain viability of the seed. Germination requirements must be known so that periodic tests can be made.

Care must be taken, as training materials regarding seed production, cleaning, storage, and maintenance often focus on making landraces more uniform, distinct and stable (usually for commercial application) which can result in the loss of valuable adaptive traits unique to local varieties.

Additionally, there is a matter of localized nature to be considered. In the upper northern hemisphere, and lower southern, one sees a seasonal change in terms of a cooler winter. Many plants go-to-seed and then go dormant. These seeds must hibernate until their respective spring season.

Open Pollination

Open pollination is an important aspect of seed saving. Plants that reproduce through natural means tend to adapt to local conditions over time, and evolve as reliable performers, particularly in their localities, known as landraces or "folk varieties."

Legality

While saving seed and even exchanging seed with other farmers for biodiversity purposes has been a traditional practice, these practices have become illegal for the plant varieties that are patented or otherwise owned by some entity (often a corporation). Under Article 28 of the World Trade Organization (WTO) Agreement on Trade-Related Aspects of Intellectual Property Rights (the TRIPS Agreement), "planting, harvesting, saving, re-planting, and exchanging seeds of patented plants, or of plants containing patented cells and genes, constitutes use" and can in some cases be prohibited by the intellectual property laws of WTO Members.

Significantly, farmers in developing countries are particularly affected by prohibitions on seed saving. There are some protections for re-use, called "farmer's privilege", in the 1991 International Union for the Protection of New Varieties of Plants (UPOV Convention), but seed exchange remains prohibited.

In the United States, the farmer's privilege to save seeds to grow subsequent crops was considered protected by the Plant Variety Protection Act of 1970. American farmers, it was thought, could sell seed up to the amount saved for replanting their own acreage.

That view came to an end in the latter part of the 20th century and early part of the 21st century, with changes in technology and law. First, in 1981 *Diamond v. Chakrabarty* established that companies may obtain patents for life-forms—originally genetically engineered unicellular bacteria. In 2002 *J.E.M. Ag Supply v. Pioneer* established that valid utility patents could be issued on sexually reproduced plants, such as seed crops (e.g., corn). In 2013 *Bowman v. Monsanto Co.* established that it was patent infringement for farmers to save crop seeds (soybeans in that case) and grow subsequent crops from them, if the seeds or plants were patented.

Seed Paper

Seed paper is a type of handmade paper that includes any number of different plant seeds. The seeds themselves can still germinate after the paper making process and they can sprout when the paper is planted in soil.

Papermakers have been producing paper including seeds in the United States since 1941, but international papermakers have practiced seed inclusion in paper for centuries. Seed paper has traditionally been handmade in smaller batches and is often made-to-order for clients.

Sprouting seed paper has enjoyed a resurgence of popularity in the United States recently. Seed paper can be used for stationery, cards, invitations, and for decorative wraps.

A wide variety of flower, vegetable, and tree seeds can also be used in seed paper for decorative effect. The seeds and flowers in the paper can create decorative effects and colors. Depending on the type of seed and the process used, different colors, thickness, and patterns can be created.

Seed paper is now used across the UK and Europe in the consumer market and the events and marketing industry. Brands are becoming aware of the engagement that can be achieved with seed paper while also aligning themselves with an environmentally friendly message.

Seed Orchard

Seed orchard in Mimizan, Landes, France

A seed orchard is an intensively-managed plantation of specifically arranged trees for the mass production of genetically improved seeds to create plants, or seeds for the establishment of new forests.

General

Seed orchards are a common method of mass-multiplication for transferring genetically improved material from breeding populations to production populations (forests) and in this sense are often referred to as "multiplication" populations. A seed orchard is often composed of grafts (vegetative copies) of selected genotypes, but seedling seed orchards also occur mainly to combine orchard with progeny testing. Seed orchards are the strong link between breeding programs and plantation establishment. They are designed and managed to produce seeds of superior genetic quality compared to those obtained from seed production areas, seed stands, or unimproved stands.

Material and Connection with Breeding Population

In first generation seed orchards, the parents usually are phenotypically-selected trees. In advanced generation seed orchards, the seed orchards are harvesting the benefits generated by tree breeding and the parents may be selected among the tested clones or families. It is efficient to synchronise the productive live cycle of the seed orchards with the cycle time of the breeding population. In the seed orchard, the trees can be arranged in a design to keep the related individuals or cloned copies apart from each other. Seed orchards are the delivery vehicle for genetic improvement programs where the trade-off between genetic gain and diversity is the most important concern. The genetic gain of seed orchard crops depends primarily on the genetic superiority of the orchard parents, the gametic contribution to the resultant seed crops, and pollen contamination from outside seed orchards.

Genetic Diversity of Seed Orchard Crops

Seed production and gene diversity is an important aspect when using improved materials like seed orchard crops. Seed orchards crops derive generally from a limited number of trees. But if it is a common wind-pollinated species much pollen will come from outside the seed orchard and widen the genetic diversity. The genetic gain of the first generation seed orchards is not great and the seed orchard progenies overlap with unimproved material. Gene diversity of the seed crops is greatly influenced by the relatedness (kinship) among orchard parents, the parental fertility variation, and the pollen contamination.

Management and Practical Examples

Seed orchards are usually managed to obtain sustainable and large crops of seeds of good quality. To achieve this, the following methods are commonly applied: orchards are established on flat surface sites with southern exposure (better conditions for orchard maintenance and for seed production), no stands of the same species in close proximity (avoid strong pollen contamination), sufficient area to produce and be mainly pollinated with their own pollen cloud, cleaning the corridors between the rows, fertilising, and supplemental pollination. The genetic quality of seed orchards can be improved by genetic thinning and selective harvesting. In plantation forestry with southern yellow pines in the United States, almost all plants originate from seed orchards and most plantations are planted in family blocks, thus the harvest from each clone is kept separate during seed processing, plant production and plantation. Recent conference proceedings about seed orchards is available on the net .

Recent Seed Orchard Research

- The optimal balance between the effective number of clones (diversity, status number, gene diversity) and genetic gain is achieved by making clonal contributions (number of ramets) proportional (linearly dependent) to the genetic value ("linear deployment"). This is dependent on several assumptions, one of them that the contribution to the seed orchard crop is proportional to the number of ramets. But the more ramets the larger the share of the pollen is lost depending on ineffective self-pollination. But even considering this, the linear deployment is a very good approximation. It was thought that increasing the gain is always accompanied by a loss in effective number of clones, but it has shown that both can be obtained in the same time by genetic thinning using the linear deployment algorithm if applied to some rather unbalanced seed orchards. Note that relatedness among clones is more critical for diversity than inbreeding.
- The clonal variation in expected seed set has been compiled for 12 adult clonal seed orchards of Scots pine. The seed set ability is not that drastic among clones as has been shown in other investigations which are probably less relevant for actual seed production of Scots pine.
- The correlations of cone set for Scots pine in a clonal archive was not well correlated with that of the same clones in seed orchards. Thus it does not seem meaningful to increase seed set by choosing clones with a good seed set.
- As supporting tree breeding make advances, new seed orchards will be genetically better than old ones. This is a relevant factor for the economic lifetime of a seed orchard. Considerations for Swedish Scots pine suggested an economic lifetime of 30 years, which is less than the current lifetime.
- Seed orchards for important wind pollinated species start to produce seeds before the seed orchard trees start to produce much pollen. Thus all or most of the pollen parents are outside the seed orchard. Calculations indicates that seed orchard seeds are still to be expected to a superior alternative to older and more mature seed orchards or stand seeds. Advantage of early seeds like absence of selfing or related matings and high diversity are positive factors in the early seeds.
- Swedish conifers orchards with tested clones could have 20-25 clones with more ramets from the better and less from the worse so effective ramet number is 15-18. Higher clone number results in unneeded loss of genetic gain. Lower clone numbers can still be better than existing alternatives. For southern pines in United States it may be optimal with half as many clones.
- When forest tree breeding proceeds to advanced generations the candidates to seed orchards will be related and the question to what degree related clones can be tolerated in seed orchards become urgent. Gene diversity seems to be a more important consideration than inbreeding. If the number of candidates have at least eight times as much diversity

(status number) as required for the seed orchard relations are not limiting and clones can be deployed as usual but restricting for half and full sibs, but if the candidate population has a lower diversity more sophisticated algorithms are needed.

Seed Testing

Seed testing is performed for a number of reasons, including research purposes or to determine if seed storage techniques are functioning. There are four tests most commonly done. The first two listed below are common for scientific research.

For commercially sold seed, all four of these tests are done in dedicated laboratories by trained and usually certified analysts. The tests are designed to evaluate the quality of the seed lot being sold.

- Germination test : Reports the percentage of seed that germinated. In commercial settings, tests are usually made in either 200 or 400 seed samples.
- Viability test (TZ test): A test for viability that involves three steps:
    - 1. preconditioning (imbibition)
    - 2. preparation and staining (sometimes cutting the seed and then soaking the seed in a 2,3,5 triphenyl tetrazolium chloride solution)
    - 3. evaluation (examining the seed for a color change in the embryo).
- Purity test: The percentage of seed described on the label that is actually found in the quantity of seed.
- Weed test: Examines a sample of seed and identifies every seed that is different from the labeled seed kind.

Seed Drill

A sowing machine which uses the seed drill concept

A seed drill is a sowing device that positions seeds in the morning and then covers them. The seed drill sows the seeds at equal distances and proper depth, ensuring that the seeds get covered with soil and are saved from being eaten by birds. Before the introduction of the seed drill, a common practice was to plant seeds by hand. Besides being wasteful, planting was very imprecise and led to a poor distribution of seeds, leading to low productivity. Jethro Tull is widely thought of as having invented the seed drill, though earlier the Sumerians used a single-tube seed drill, and the Chinese used a multi-tube seed drill. The use of a seed drill can improve the ratio of crop yield (seeds harvested per seed planted) by as much as nine times. In short, the seed drill can be described as a modern agricultural implement used for sowing seeds.

Design

In older methods of planting, a field is initially prepared with a plough to a series of linear cuts known as *furrows*. The field is then seeded by throwing the seeds over the field, a method known as *manual broadcasting*. In this method there will be a right depth nor proper distance. Seeds that landed in the furrows had better protection from the elements, and natural erosion or manual raking would preferentially cover them while leaving some exposed. The result was a field planted roughly in rows, but having a large number of plants outside the furrow lanes.

There are several downsides to this approach. The most obvious is that seeds that land outside the furrows will not have the growth shown by the plants sown in the furrow, since they are too shallow on the soil. Because of this, they are lost to the elements. Much of the seed remained on the surface where it never germinated or germinated prematurely, only to be killed by frost. On the surface, it was also vulnerable to being eaten by birds or carried away on the wind.

Since the furrows represent only a portion of the field's area, and broadcasting distributes seeds fairly evenly, this results in considerable wastage of seeds. Less obvious are the effects of overseeding; all crops grow best at a certain density, which varies depending on the soil and weather conditions. Additional seeding above this limit will actually reduce crop yields, in spite of more plants being sown, as there will be competition among the plants for the minerals, water and the soil available. Another reason is that the mineral resources of the soil will also deplete at a much faster rate, thereby directly affecting the growth of the plants.

The invention of the seed drill dramatically improved germination. The seed drill employed a series of runners spaced at the same distance as the ploughed furrows. These runners, or drills, opened the furrow to a uniform depth before the seed was dropped. Behind the drills were a series of presses, metal discs which cut down the sides of the trench into which the seeds had been planted, covering them over.

This innovation permitted farmers to have precise control over the depth at which seeds were planted. This greater measure of control meant that fewer seeds germinated early or late, and that seeds were able to take optimum advantage of available soil moisture in a prepared seed bed. The result was that farmers were able to use less seed, and at the same time experience larger yields than under the broadcast methods.

History

Chinese double-tube seed drill, published by Song Yingxing in the *Tiangong Kaiwu* encyclopedia of 1637.

While the Babylonians used primitive seed drills around 1500 BCE, the invention never reached Europe. Multi-tube iron seed drills were invented by the Chinese in the 2nd century BCE. This multi-tube seed drill has been credited with giving China an efficient food production system that allowed it to support its large population for millennia. This multi-tube seed drill may have been introduced into Europe following contacts with China.

The first known European seed drill was attributed to Camillo Torello and patented by the Venetian Senate in 1566. A seed drill was described in detail by Tadeo Cavalina of Bologna in 1602. In England, the seed drill was further refined by Jethro Tull in 1701 in the Agricultural Revolution. However, seed drills of this and successive types were both expensive and unreliable, as well as fragile. Seed drills would not come into widespread use in Europe until the mid-19th century. Early drills were small enough to be pulled by a single horse, and many of these remained in use into the 1930s. The availability of steam, and later gasoline tractors, however, saw the development of larger and more efficient drills that allowed farmers to seed ever larger tracts in a single day.

Recent improvements to drills allow seed-drilling without prior tilling. This means that soils subject to erosion or moisture loss are protected until the seed germinates and grows enough to keep the soil in place. This also helps prevent soil loss by avoiding erosion after tilling. The development of the press drill was one of the major innovations in pre-1900 farming technology.

Uses

Drilling is the term used for the mechanised sowing of an agricultural crop. Traditionally, a seed drill used to consist of a hopper filled with seeds arranged above a series of tubes that can be set at selected distances from each other to allow optimum growth of the resulting plants. Seeds are spaced out using fluted paddles which rotate using a geared drive from one of the drill's land wheels—seed rate is altered by changing gear ratios. Most modern drills use air to convey seed in plastic tubes from the seed hopper to the coulters—it is an arrangement which allows seed drills to be much wider than the seed hopper—as much as 12 m wide in some cases. The seed is metered mechanically into an air stream created by a hydraulically powered on-board fan and conveyed initially to a distribution head which sub-divides the seed into the pipes taking the seed to the individual coulters.

1902 model 12-run seed drill produced by Monitor Manufacturing Company, Minneapolis, Minnesota.

The seed drill allows farmers to sow seeds in well-spaced rows at specific depths at a specific seed rate; each tube creates a hole of a specific depth, drops in one or more seeds, and covers it over. This invention gives farmers much greater control over the depth that the seed is planted and the ability to cover the seeds without back-tracking. This greater control means that seeds germinate consistently and in good soil. The result is an increased rate of germination, and a much-improved crop yield (up to eight times).

A further important consideration is weed control. Broadcast seeding results in a random array of growing crops, making it difficult to control weeds using any method other than hand weeding. A field planted using a seed drill is much more uniform, typically in rows, allowing weeding with the hoe during the course of the growing season. Weeding by hand is laborious and poor weeding limits yield.

The ground would have to be plowed and harrowed. The plow would dig up the earth and the harrow would smooth the soil and break up any clumps. The drill would be set for the size of seed used. Then the grain would be put in the hopper on top and then follow along behind it while the seed drill spaced and planted the seed. This system is still used today but it has been modified and updated so a farmer can plant many rows of seed at the same time.

A seed drill can be pulled across the field using bullocks or a tractor. Seeds sown using a seed drill are distributed evenly and placed at the correct depth in the soil.

References

- Black, Michael H.; Halmer, Peter (2006). The encyclopedia of seeds: science, technology and uses. Wallingford, UK: CABI. p. 224. ISBN 978-0-85199-723-0.
- Joseph Needham; Gwei-Djen Lu; Ling Wang (1987). Science and civilisation in China. Cambridge University Press. pp. 48–50. ISBN 978-0-521-30358-3.
- Davenport, Paul (20 April 2009). "Photo-Radar Van Driver Shot to Death". MyFoxPhoenix.com. Retrieved 19 January 2014.
- "Speed cameras will stay – but no more maintenance | News". This is Gloucestershire. 3 August 2010. Retrieved 24 May 2012.

- Humphrey, Tom. "Speed cameras catch police chief's interest » Knoxville News Sentinel". Knoxnews.com. Retrieved 24 May 2012.
- "Redflex challenges private investigators ruling – The Independent Weekly". Theind.com. 19 May 2008. Retrieved 24 May 2012.
- Day, M. V.; Ross, M. (2011). "The value of remorse: How drivers' responses to police predict fines for speeding". Law and Human Behavior. 35 (3): 221–234. doi:10.1007/s10979-010-9234-4. PMID 20556494.
- Alexandra Smith (31 March 2011). "Greater Leeway for Speeding Drivers". The Age. Australia. Retrieved 16 June 2011.
- "'Cash cow' speed cameras raise $350m". The Sydney Morning Herald. Fairfax Media. 2 June 2011. Retrieved 27 June 2011.
- Report finds speed cameras are an effective safety tool Media Release The Hon Duncan Gay MLC Minister for Roads and Ports 27 July 2011
- "PARLIAMENT is facing a renewed push for a top-level inquiry into the use of speed cameras". Adelaide Now. News Limited. 27 June 2011. Retrieved 27 June 2011.
- "Speeding and disobedient drivers deliver a windfall for government coffers". The Herald Sun. Herald and Weekly Times. 11 June 2011. Retrieved 19 June 2011.

4

Seed Bank: An Integrated Study

Seed banks are banks that store seeds; it is done in order to preserve genetic diversity of seeds. Soil seed bank and Svalbard global seed vault are some of the topics discussed in this section. This section will provide an integrated understanding of seed bank.

Seed Bank

Seedbank at the USDA Western Regional Plant Introduction Station

A seed bank (also seedbank or seeds bank) stores seeds to preserve genetic diversity; hence it is a type of gene bank. There are many reasons to store seeds. One reason is to have on-hand the genes that plant breeders need to increase yield, disease resistance, drought tolerance, nutritional quality, etc. of plants used in agriculture (i.e., crops or domesticated species). Another reason is to forestall loss of genetic diversity in rare or imperiled plant species in an effort to conserve biodiversity ex situ. Many plants that were used centuries ago by humans are used less frequently now and seed banks offer a way to preserve that historical and cultural value. Collections of seeds stored at constant low temperature and moisture guard against loss of genetic resources that are otherwise maintained in situ or in field collections. These alternative 'living' collections can be damaged by natural disasters, outbreaks of disease or war. Seed banks are considered seed libraries and contain valuable information about evolved strategies to combat plant stress or produce novel products. The work of seed banks spans decades and even centuries. Most seed banks are publicly funded and seeds are usually available for research that benefits the public.

Seed banks store seeds to keep them viable. Distribution of seeds from seed banks is a form of swapping seeds

The Global Seed Vault is situated in Svalbard, midway between Norway and the Arctic.

Storage Conditions and Regeneration

Seeds are living creatures and keeping them viable over the long term requires adjusting storage moisture and temperature appropriately. As they mature on the mother plant, many seeds attain an innate ability to survive drying. Survival of these so-called 'orthodox' seeds can be extended by dry, low temperature storage. The level of dryness and coldness depends mostly on the longevity that is required and the investment in infrastructure that is affordable. Practical guidelines from a US scientist in the 1950s and 1960s, James Harrington, are known as 'Thumb Rules.' The 'Hundreds Rule' guides that the sum of relative humidity and temperature (in Fahrenheit) should be less than 100 for the sample to survive 5 years. Another rule is that reduction of water content by 1% or temperature by 10 degrees Fahrenheit will double the seed life span. Research from the 1990s showed that there is a limit to the beneficial effect of drying or cooling, so it must not be overdone.

Understanding the effect of water content and temperature on seed longevity, the Food and Agriculture division of the United Nations and a consultancy group called Bioversity International developed a set of standards for international seed banks to preserve seed longevity. The document advocates drying seeds to about 20% relative humidity, sealing seeds in high quality moisture-proof containers, and storing seeds at -20 degrees Celsius. These conditions are frequently referred to as 'conventional' storage protocols. Seeds from our most important species - corn, wheat, rice, soybean, pea, tomato, broccoli, melon, sunflower, etc.—can be stored in this way. However, there are many species that produce seeds that do not survive the drying or low temperature of conventional storage protocols. These species must be stored cryogenically. Seeds of citrus, coffee, avocado, cocoa, coconut, papaya, oak, walnut and willow are a few examples of species that should be preserved cryogenically.

Like everything, seeds eventually degrade with time. It is hard to predict when seeds lose viability and so most reputable seed banks monitor germination potential during storage. When seed germination percentage decreases below a prescribed amount, the seeds need to be replanted and fresh seeds collected for another round of long-term storage.

Challenges

- Knowing what to store in a seed bank is the greatest challenge. Collections must be relevant and that means they must provide useful genetic diversity that is accessible to the public. Collections must also be efficient and that means they mustn't duplicate materials already in collections.
- Keeping seeds alive for hundreds of years is the next biggest challenge. Orthodox seeds are amenable to 'conventional' storage protocols but there are many seed types that must be stored using nonconventional methods. Technology for these methods is rapidly advancing; Local institutional infrastructure may be lacking.

Alternatives

In-situ conservation of seed-producing plant species is another conservation strategy. In-situ conservation involves the creation of National Parks, National Forests, and National Wildlife Refuges as a way of preserving the natural habitat of the targeted seed-producing organisms. In-situ conservation of agricultural resources is performed on-farm. This also allows the plants to continue to evolve with their environment through natural selection.

An arboretum stores trees by planting them at a protected site.

A less expensive, community-supported seed library can save local genetic material.

The phenomenon of seeds remaining dormant within the soil is well known and documented (Hills and Morris 1992). Detailed information on the role of such "seed banks" in northern Ontario, however, is extremely limited, and research is required to determine the species and abundance of seeds in the soil across a range of forest types, as well as to determine the function of the seed bank in post-disturbance vegetation dynamics. Comparison tables of seed density and diversity are presented for the boreal and deciduous forest types and the research that has been conducted is discussed. This review includes detailed discussions of: (1) seed bank dynamics, (2) physiology of seeds in a seed bank, (3) boreal and deciduous forest seed banks, (4) seed bank dynamics and succession, and (5) recommendations for initiating a seed bank study in northern Ontario.

Longevity

Seeds may be viable for hundreds and even thousands of years. The oldest carbon-14-dated seed that has grown into a viable plant was a Judean date palm seed about 2,000 years old, recovered from excavations at Herod the Great's palace in Israel.

In February 2012, Russian scientists announced they had regenerated a narrow leaf campion (Silene stenophylla) from a 32,000-year-old seed. The seed was found in a burrow 124 feet under Siberian permafrost along with 800,000 other seeds. Seed tissue was grown in test tubes until it could be transplanted to soil. This exemplifies the long-term viability of DNA under proper conditions.

Facilities

Plant tissue cultures being grown at a USDA seed bank, the National Center for Genetic Resources Preservation.

There are about 6 million accessions, or samples of a particular population, stored as seeds in about 1,300 genebanks throughout the world as of 2006. This amount represents a small fraction of the world's biodiversity, and many regions of the world have not been fully explored.

- The Svalbard Global Seed Vault has been built inside a sandstone mountain in a man-made tunnel on the frozen Norwegian island of Spitsbergen, which is part of the Svalbard archipelago, about 1,307 kilometres (812 mi) from the North Pole. It is designed to survive catastrophes such as nuclear war and world war. It is operated by the Global Crop Diversity Trust. The area's permafrost will keep the vault below the freezing point of water, and the seeds are protected by 1-metre thick walls of steel-reinforced concrete. There are two airlocks and two blast-proof doors. The vault accepted the first seeds on 26 February 2008.

- The Millennium Seed Bank housed at the Wellcome Trust Millennium Building (WTMB), located in the grounds of Wakehurst Place in West Sussex, near London, in England, UK. It is the largest seed bank in the world (longterm, at least 100 times bigger than Svalbard Global Seed Vault), providing space for the storage of billions of seed samples in a nuclear bomb proof multi-story underground vault. Its ultimate aim being to store every plant species possible, it reached its first milestone of 10% in 2009, with the next 25% milestone aimed to be reached by 2020. Importantly they also distribute seeds to other key locations around the world, do germination tests on each species every 10 years, and other important research.

- The former NSW Seedbank focuses on native Australian flora, especially NSW threatened species. The project was established in 1986 as an integral part of The Australian Botanic Gardens, Mount Annan. The NSW Seedbank hasdcollaborated with the Millennium Seed Bank since 2003. The seed bank has since been replaced as part of a major upgrade by the Australian PlantBank.

- Nikolai Vavilov (1887–1943) was a Russian geneticist and botanist who, through botanic-agronomic expeditions, collected seeds from all over the world. He set up one of the first seed banks, in Leningrad (now St Petersburg), which survived the 28-month Siege of Leningrad in World War II. It is now known as the Vavilov Institute of Plant Industry. Several botanists starved to death rather than eat the collected seeds.

- The BBA (Beej Bachao Andolan — Save the Seeds movement) began in the late 1980s in Uttarakhand, India, led by Vijay Jardhari. Seed banks were created to store native varieties of seeds.

- National Center for Genetic Resources Preservation, Fort Collins, Colorado, United States

- Desert Legume Program (DELEP) focuses on wild species of plants in the legume family (Fabaceae), specifically legumes from dry regions around the world. The DELEP seed bank currently has over 3600 seed collections representing nearly 1400 species of arid land legumes originating in 65 countries on six continents. It is backed up (at least in part) in National Center for Genetic Resources Preservation, and in the Svalbard Global Seed Vault. The DELEP seed bank is an accredited collection of the North American Plant Conservation Consortium.

Soil Seed Bank

The soil seed bank is the natural storage of seeds, often dormant, within the soil of most ecosystems. The study of soil seed banks started in 1859 when Charles Darwin observed the emergence of seedlings using soil samples from the bottom of a lake. The first scientific paper on the subject was published in 1882 and reported on the occurrence of seeds at different soil depths. Weed seed banks have been studied intensely in agricultural science because of their important economic impacts; other fields interested in soil seed banks include forest regeneration and restoration ecology.

Background

Many taxa have been classified according to the longevity of their seeds in the soil seed bank. Seeds of *transient* species remain viable in the soil seed bank only to the next opportunity to germinate, while seeds of *persistent* species can survive longer than the next opportunity—often much longer than one year. Species with seeds that remain viable in the soil longer than five years form the *long-term* persistent seed bank, while species whose seeds generally germinate or die within one to five years are called *short-term* persistent. A typical long-term persistent species is *Chenopodium album* (Lambsquarters); its seeds commonly remain viable in the soil for up to 40 years and in rare situations perhaps as long as 1,600 years. A species forming no soil seed bank at all (except the dry season between ripening and the first autumnal rains) is *Agrostemma githago* (Corncockle), which is a formerly widespread cereal weed.

Seed Longevity

Dried lotus seeds.

Longevity of seeds is very variable and depends on many factors; few species exceed 100 years. In typical soils the longevity of seeds can range from nearly zero (germinating immediately when reaching the soil or even before) to several hundred years. Some of the oldest still-viable seeds were those of Lotus (*Nelumbo nucifera*) found buried in the soil of a pond; these seeds were estimated by carbon dating to be around 1,200 years old.

One of the longest-running soil seed viability trials was started in Michigan in 1879 by James Beal. The experiment involved the burying of 20 bottles holding 50 seeds from 21 species. Every five years, a bottle from every species was retrieved and germinated on a tray of sterilized soil which was kept in a growth chamber. Later, after responsibility for managing the experiment was delegated to caretakers, the time frame between each retrieval became longer. In 1980, more than 100 years after the trial was started, seeds of only three species were observed to germinate: moth mullein (*Verbascum blattaria*), common mullein (*Verbascum thapsus*) and common mallow (*Malva neglecta*).

Environmental Significance

Soil seed banks play an important role in the natural environment of many ecosystems. For example, the rapid re-vegetation of sites disturbed by wildfire, catastrophic weather, agricultural operations, and timber harvesting is largely due to the soil seed bank. Forest ecosystems and wetlands contain a number of specialized plant species forming persistent soil seed banks.

Before the advent of herbicides a good example of a persistent seed bank species, Papaver rhoeas sometimes was so abundant in agricultural fields in Europe that it could be mistaken for a crop.

The absence of a soil seed bank impedes the establishment of vegetation during primary succession, while presence of a well-stocked soil seed bank permits rapid development of species-rich ecosystems during secondary succession.

Population Densities and Diversity

The mortality of seeds in the soil is one of the key factors for the persistence and density fluctuations of plant populations, especially for annual plants. Studies on the genetic structure of *Androsace septentrionalis* populations in the seed bank compared to those of established plants showed that diversity within populations is higher below ground than above ground.

There are indications that mutations are more important for species forming a persistent seed bank compared to those with only transient seeds. The increase of species richness in a plant community due to a species-rich and abundant soil seed bank is known as the *storage effect*.

Species of *Striga* (witchweed) are known to leave some of the highest seed densities in the soil compared to other plant genera; this is a major factor that aids their invasive potential. Each plant has the capability to produce between 90,000 and 450,000 seeds, although a majority of these seeds are not viable. It has been estimated that only two witchweeds would produce enough seeds required to refill a seed bank after seasonal losses.

Associated Ecosystem Processes

The term soil diaspore bank can be used to include non-flowering plants such as ferns and bryophytes.

In addition to seeds, perennial plants have vegetative propagules to facilitate forming new plants, migration into new ground, or reestablishment after being top-killed. These propagules are collectively called the 'soil bud bank', and include dormant and adventitious buds on stolons, rhizomes, and bulbs.

Svalbard Global Seed Vault

The Svalbard Global Seed Vault (Norwegian: *Svalbard globale frøhvelv*) is a secure seed bank on the Norwegian island of Spitsbergen near Longyearbyen in the remote Arctic Svalbard archipelago, about 1,300 kilometres (810 mi) from the North Pole. Conservationist Cary Fowler, in association with the Consultative Group on International Agricultural Research (CGIAR), started the vault to preserve a wide variety of plant seeds that are duplicate samples, or "spare" copies, of seeds held in gene banks worldwide. The seed vault is an attempt to insure against the loss of seeds in other genebanks during large-scale regional or global crises. The seed vault is managed under terms spelled out in a tripartite agreement between the Norwegian government, the Global Crop Diversity Trust (GCDT) and the Nordic Genetic Resource Center (NordGen).

The Norwegian government entirely funded the vault's approximately NOK 45 million (US$9 million) construction. Storing seeds in the vault is free to end users, with Norway and the Global Crop Diversity Trust paying for operational costs. Primary funding for the Trust comes from organisations such as the Bill & Melinda Gates Foundation and from various governments worldwide.

History

The Nordic Gene Bank (NGB) has, since 1984, stored backup Nordic plant germplasm via frozen seeds in an abandoned coal mine at Svalbard, over the years depositing more than 10,000 seed samples of more than 2,000 cultivars for 300 different species. The Nordic collection has for years duplicated seed samples from the Southern African Development Community. Both the Nordic and African collections have been transferred to the new Svalbard Global Seed Vault facility. On 1 January 2008 the Nordic Gene Bank was integrated with NordGen.

Construction

Entrance to the Vault

Norway, Sweden, Finland, Denmark, and Iceland's prime ministers ceremonially laid "the first stone" on 19 June 2006.

The seedbank is 120 metres (390 ft) inside a sandstone mountain on Spitsbergen Island, and employs robust security systems. Seeds are packaged in special three-ply foil packets and heat sealed to exclude moisture. The facility is managed by the Nordic Genetic Resource Center, though there are no permanent staff on-site.

Spitsbergen was considered ideal because it lacked tectonic activity and had permafrost, which aids preservation. Its being 130 metres (430 ft) above sea level will keep the site dry even if the ice caps melt. Locally mined coal provides power for refrigeration units that further cool the seeds to the internationally recommended standard of −18 °C (−0.4 °F). If the equipment fails, at least several weeks will elapse before the facility rises to the surrounding sandstone bedrock's temperature of −3 °C (27 °F).

A feasibility study prior to construction determined that the vault could, for hundreds of years, preserve most major food crops' seeds. Some, including those of important grains, could survive far longer—possibly thousands of years.

The Svalbard Global Seed Vault officially opened on 26 February 2008. Approximately 1.5 million distinct seed samples of agricultural crops are thought to exist. The variety and volume of seeds stored will depend on the number of countries participating – the facility has a capacity to conserve 4.5 million. The first seeds arrived in January 2008. Five percent of the seeds in the vault, about 18,000 samples with 500 seeds each, come from the Centre for Genetic Resources of the Netherlands (CGN), part of Wageningen University, Netherlands. By 2013, approximately one-third of the genera diversity stored in gene banks globally was represented at the Seed Vault.

Public Art

Illuminated art installation above the entrance to the Vault

Running the length of the facility's roof and down the front face to the entryway is an illuminated artwork named *Perpetual Repercussion* by Norwegian artist Dyveke Sanne that marks the location of the vault from a distance. In Norway, government-funded construction projects exceeding a certain cost must include artwork. KORO, the Norwegian State agency overseeing art in public spaces, engaged the artist to install lighting that highlights the importance and qualities of Arctic light. The roof and vault entrance are filled with highly reflective stainless steel, mirrors, and prisms. The installation reflects polar light in the summer months, while in the winter, a network of 200 fibre-optic cables gives the piece a muted greenish-turquoise and white light.

Mission

The Svalbard Global Seed Vault's mission is to provide a safety net against accidental loss of diversity in traditional genebanks. While the popular press has emphasized its possible utility in the event of a major regional or global catastrophe, it will be more frequently accessed when genebanks lose samples due to mismanagement, accident, equipment failures, funding cuts, and natural disasters. These events occur with some regularity. War and civil strife have a history of destroying some genebanks. The national seed bank of the Philippines was damaged by flooding and later destroyed by a fire; the seed banks of Afghanistan and Iraq have been lost completely. According to *The Economist*, "the Svalbard vault is a backup for the world's 1,750 seed banks, storehouses of agricultural biodiversity." By the request of Norwegian government, no genetically modified seeds are stored at the vault.

Access to Seeds

Vault seed samples are copies of samples stored in the depositing genebanks. Researchers, plant breeders, and other groups wishing to access seed samples cannot do so through the seed vault; they must instead request samples from the depositing genebanks. The samples stored in the genebanks will, in most cases, be accessible in accordance with the terms and conditions of the International Treaty on Plant Genetic Resources for Food and Agriculture, approved by 118 countries or parties.

The seed vault functions like a safe deposit box in a bank. The bank owns the building and the depositor owns the contents of his or her box. The Government of Norway owns the facility and the depositing genebanks own the seeds they send. The deposit of samples in Svalbard does not constitute a legal transfer of genetic resources. In genebank terminology this is called a "black box" arrangement. Each depositor signs a Deposit Agreement with NordGen, acting on behalf of Norway. The Agreement makes clear that Norway does not claim ownership over the deposited samples and that ownership remains with the depositor, who has the sole right of access to those materials in the seed vault. No one has access to anyone else's seeds from the seed vault. The database of samples and depositors is maintained by NordGen.

The Syrian Civil War caused another seed bank, the International Center for Agricultural Research in the Dry Areas, to move its headquarters from Aleppo to Beirut. Due to difficulties by ICARDA in transferring its collection, in 2015 the Svalbard Vault authorized the first withdrawal of seeds in its history.

Seed Storage

Seed storage containers on metal shelving inside the vault

The seeds are stored in sealed three-ply foil packages, then placed into plastic tote containers on metal shelving racks. The storage rooms are kept at −18 °C (−0.4 °F). The low temperature and limited access to oxygen will ensure low metabolic activity and delay seed aging. The permafrost surrounding the facility will help maintain the low temperature of the seeds should the electricity supply fail.

Global Crop Diversity Trust

The Global Crop Diversity Trust (GCDT) has played a key role in the planning of the seed vault and is coordinating shipments of seed samples to the Vault in conjunction with the Nordic Genetic Resource Center. The Trust will provide most of the annual operating costs for the facility, and has set aside endowment funds to do so, while the Norwegian government will finance upkeep of the structure itself. With support from the Bill & Melinda Gates Foundation and other donors, the GCDT is assisting selected genebanks in developing countries as well as the international agricultural research centers in packaging and shipping seeds to the seed vault. An International Advisory Council is being established to provide guidance and advice. It will include representatives from the FAO, the CGIAR, the International Treaty on Plant Genetic Resources and other institutions.

First Anniversary Deposits

As part of the vault's first anniversary, more than 90,000 food crop seed samples were placed into storage, bringing the total number of seed samples to 400,000. Among the new seeds includes 32 varieties of potatoes from Ireland's national gene banks and 20,000 new samples from the U.S. Agricultural Research Service. Other seed samples came from Canada and Switzerland, as well as international seed researchers from Colombia, Mexico and Syria. This 4-tonne (3.9-long-ton; 4.4-short-ton) shipment brought the total number of seeds stored in the vault to over 20 million. As of this anniversary, the vault contained samples from approximately one-third of the world's most important food crop varieties. Also part of the anniversary, experts on food production and climate change met for a three-day conference in Longyearbyen.

Awards and Honors

Svalbard Global Seed Vault ranked at No. 6 on *Time*'s Best Inventions of 2008. It was awarded the Norwegian Lighting Prize for 2009.

Capacity

A seed sample consists of around 500 seeds sealed in an airtight aluminum bag, and the facility has a storage capacity of 4.5 million seed samples.

Year	Species	Total Samples	Ref.
2008		187,000+	
2010		500,000	
2013 (Feb)		774,601	
2014		820,619	
2015	4,000	840,000	

International Connections

Japanese sculptor Mitsuaki Tanabe (田辺光彰) presented a work to the vault named "The Seed 2009 / Momi In-Situ Conservation". In 2010 a delegation of seven U.S. congressmen handed over a number of different varieties of chili pepper.

Millennium Seed Bank Partnership

Millennium Seed Bank building

The Millennium Seed Bank Partnership (MSBP or MSB), formerly known as the Millennium Seed Bank Project, is an international conservation project coordinated by the Royal Botanic Gardens, Kew. After being awarded a Millennium Commission grant in 1995, the project commenced in 1996, and is now housed in the Wellcome Trust Millennium Building situated in the grounds of Wakehurst Place, West Sussex. Its purpose is to provide an "insurance policy" against the extinction of plants in the wild by storing seeds for future use. The storage facilities consist of large underground frozen vaults preserving the world's largest wild-plant seedbank or collection of seeds from wild species. The project had been started by Dr Peter Thompson and run by Paul Smith after the departure of Roger Smith. Roger Smith was awarded the OBE in 2000, in the Queen's New Year Honours for services to the Project.

The central visitor hall

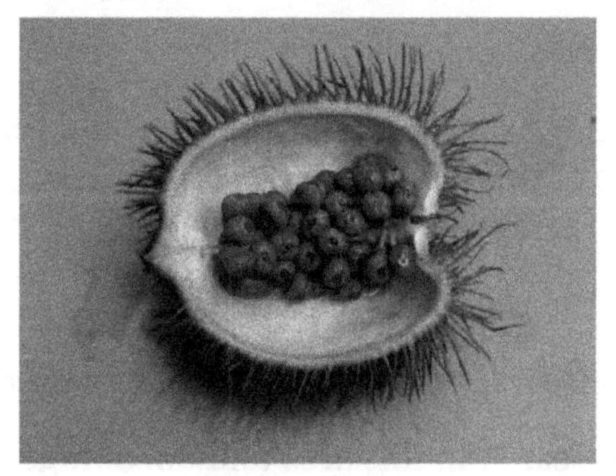

Bixa orellana seeds

In collaboration with other biodiversity projects around the world expeditions are sent to collect seeds from dryland plants. Where possible, collections are kept in the country of origin with duplicates being sent to the Millennium Seed Bank Project for storage. Major partnerships exist on all the continents, enabling the countries involved to meet international objectives such as the Global Strategy for Plant Conservation and the Millennium Development Goals of the United Nations Environment Programme.

Ravenala madagascariensis seeds

The seed bank at Kew has gone through many iterations. The Kew Seed Bank facility, set up by Peter Thompson in 1980, preceded the MSBP and was headed by Roger Smith from 1980 to 2005. From 2005, Paul Smith took over as head of the MSBP. The Wellcome Trust Millennium Seed Bank building was designed by the firm Stanton WIlliams and opened by Prince Charles in 2000. The laboratories and offices are in two wings flanking a wide space open to visitors housing an exhibition, and also allowing them to watch the work of cleaning and preparing seeds for storage through the large windows of the work areas. There is also a view down to the entrance to the underground vaults where the seeds are stored at - 20 °C. In 2001, the international programme of the MSBP was launched.

In April 2007, it banked its billionth seed, the *Oxytenanthera abyssinica*, a type of African bamboo. In October 2009, it reached its 10% goal of banking all the world's wild plant species by adding *Musa itinerans*, a wild banana, to its seed vault. As estimates for the number of seed bearing plant species have increased, 34,088 wild plant species and 1,980,405,036 seeds in storage as of June 2015 represent over 13% of the world's wild plant species.

Project Aims

The main aims of the project are to:

- Collect the seeds from 75,000 species of plants by 2020, representing 25% of known flora. This is the second phase of this goal, with the original partnership goal of banking 10% of known flora by 2010 was achieved in October 2009.
- Collect seeds from all of the UK's native flora.
- Further research into conservation and preservation of seeds and plants.
- Act as a focal point for research in this area and encourage public interest and support.

International Partnerships

Partnerships exist in Australia, Mexico, Chile, Kenya, China, United States, Jordan, Mali, Malawi,

Madagascar, Burkina Faso, Botswana, Tanzania, Saudi Arabia, Lebanon and South Africa. Australia is particularly significant as its flora constitutes 15% of the world's total of species, with 22% of them identified as under threat of extinction.

Preservation of Seeds

A placement student cleaning Pilosella officinarum at the Millennium Seed Bank. The procedure is being carried out in a dust hood.

Seed collections arrive at the MSBP in varying states, sometimes attached to fruits, sometimes clean. The collections usually also include a voucher specimen that can be used to identify the plant. The collections are immediately moved to a dry room until processing can be conducted where the seeds are cleaned of debris and other plant material, X-rayed, counted, and banked at -20°C. Seeds are banked in hermetically sealed glass containers along with Silica gel packets impregnated with indicator compounds that change colour if moisture seeps into the collection. Seeds are tested for viability with a germination test shortly after banking and then at regular 10 year intervals. If seed collections are low, re-harvesting from the wild is always the preferred option.

Seed Distribution

When seeds are required for research purposes they can be requested from the MSBP's seedlist. If it has the legal permission to do so the MSB can then provide up to 60 seeds for free, to bona fide, non-commercial organisations for the purposes of research, restoration, and re-introduction. All seeds provided to institutions are on a non-profit mutual benefit basis. The MSB also operates the UK Native Seed Hub which aims to improve the resilience of the UK's ecological networks by providing high quality UK native seeds to conservation and restoration groups.

References

- J. Derek Bewley, Michael Black and Peter Halmer (2006). The Encyclopedia of Seeds: Science, Technology and Uses. CABI. pp. 14–15. ISBN 978-0851997230.
- Ross, Merrill A.; Lembi, Carole A. (2008). Applied Weed Science: Including the Ecology and Management of Invasive Plants. Prentice Hall. p. 22. ISBN 978-0135028148.

- Jack Dekker (1997). “The Soil Seed Bank”. Agronomy Department, Iowa State University. Retrieved 10 December 2015.
- Frank W. Telewski. “Research & Teaching”. Department of Plant Biology, Michigan State University. Retrieved 10 December 2015.
- Doyle, Alister (21 September 2015). “Syrian war spurs first withdrawal from doomsday Arctic seed vault”. Reuters. Retrieved 22 September 2015.
- Robins-Early, Nick (22 September 2015). “Syrian War Causes The Global Doomsday Seed Vault’s First Withdrawal”. The Huffington Post. Retrieved 24 September 2015.
- Daniel M. Joel, Jonathan Gressel, Lytton J. Musselman (2013). Parasitic Orobanchaceae: Parasitic Mechanisms and Control Strategies. Springer Science & Business Media. p. 394.
- “Svalbard Global Seed Vault: Awarded the Norwegian Lighting Prize for 2009”. Regjeringen.no. 4 November 2009. Retrieved 15 April 2012.
- MacDougall, Ian (12 March 2010). “Norway Doomsday Seed Vault Hits 1/2 Million Mark”. Associated Press via U.S. News & World Report. Retrieved 3 July 2011.
- “Svalbard Global Seed Vault: Management and Operations”. Regjeringen.no. Archived from the original on 27 March 2010. Retrieved 3 July 2011.

5

Genetic Technology in Seed Science

Genetically modified crops are crops that have been altered with the help of genetic engineering techniques. It is mainly done with the purpose of adding a trait to the plant. Genetically modified soybean and genetically modified maize have also been explained in the section. The topics discussed in the chapter are of great importance to broaden the existing knowledge on seed science.

Genetically Modified Crops

Genetically modified crops (GMCs, GM crops, or biotech crops) are plants used in agriculture, the DNA of which has been modified using genetic engineering techniques. In most cases, the aim is to introduce a new trait to the plant which does not occur naturally in the species. Examples in food crops include resistance to certain pests, diseases, or environmental conditions, reduction of spoilage, or resistance to chemical treatments (e.g. resistance to a herbicide), or improving the nutrient profile of the crop. Examples in non-food crops include production of pharmaceutical agents, biofuels, and other industrially useful goods, as well as for bioremediation.

Farmers have widely adopted GM technology. Between 1996 and 2015, the total surface area of land cultivated with GM crops increased by a factor of 100, from 17,000 km^2 (4.2 million acres) to 1,797,000 km^2 (444 million acres). 10% of the world's arable land was planted with GM crops in 2010. In the US, by 2014, 94% of the planted area of soybeans, 96% of cotton and 93% of corn were genetically modified varieties. Use of GM crops expanded rapidly in developing countries, with about 18 million farmers growing 54% of worldwide GM crops by 2013. A 2014 meta-analysis concluded that GM technology adoption had reduced chemical pesticide use by 37%, increased crop yields by 22%, and increased farmer profits by 68%. This reduction in pesticide use has been ecologically beneficial, but benefits may be reduced by overuse. Yield gains and pesticide reductions are larger for insect-resistant crops than for herbicide-tolerant crops. Yield and profit gains are higher in developing countries than in developed countries.

There is a scientific consensus that currently available food derived from GM crops poses no greater risk to human health than conventional food, but that each GM food needs to be tested on a case-by-case basis before introduction. Nonetheless, members of the public are much less likely than scientists to perceive GM foods as safe. The legal and regulatory status of GM foods varies by country, with some nations banning or restricting them, and others permitting them with widely differing degrees of regulation.

However, opponents have objected to GM crops on several grounds, including environmental concerns, whether food produced from GM crops is safe, whether GM crops are needed to address the world's food needs, and concerns raised by the fact these organisms are subject to intellectual property law.

Gene Transfer in Nature and Traditional Agriculture

DNA transfers naturally between organisms. Several natural mechanisms allow gene flow across species. These occur in nature on a large scale – for example, it is one mechanism for the development of antibiotic resistance in bacteria. This is facilitated by transposons, retrotransposons, proviruses and other mobile genetic elements that naturally translocate DNA to new loci in a genome. Movement occurs over an evolutionary time scale.

The introduction of foreign germplasm into crops has been achieved by traditional crop breeders by overcoming species barriers. A hybrid cereal grain was created in 1875, by crossing wheat and rye. Since then important traits including dwarfing genes and rust resistance have been introduced. Plant tissue culture and deliberate mutations have enabled humans to alter the makeup of plant genomes.

History

The first genetically modified crop plant was produced in 1982, an antibiotic-resistant tobacco plant. The first field trials occurred in France and the USA in 1986, when tobacco plants were engineered for herbicide resistance. In 1987, Plant Genetic Systems (Ghent, Belgium), founded by Marc Van Montagu and Jeff Schell, was the first company to genetically engineer insect-resistant (tobacco) plants by incorporating genes that produced insecticidal proteins from *Bacillus thuringiensis* (Bt).

The People's Republic of China was the first country to allow commercialized transgenic plants, introducing a virus-resistant tobacco in 1992, which was withdrawn in 1997. The first genetically modified crop approved for sale in the U.S., in 1994, was the *FlavrSavr* tomato. It had a longer shelf life, because it took longer to soften after ripening. In 1994, the European Union approved tobacco engineered to be resistant to the herbicide bromoxynil, making it the first commercially genetically engineered crop marketed in Europe.

In 1995, Bt Potato was approved by the US Environmental Protection Agency, making it the country's first pesticide producing crop. In 1995 canola with modified oil composition (Calgene), Bt maize (Ciba-Geigy), bromoxynil-resistant cotton (Calgene), Bt cotton (Monsanto), glyphosate-resistant soybeans (Monsanto), virus-resistant squash (Asgrow), and additional delayed ripening tomatoes (DNAP, Zeneca/Peto, and Monsanto) were approved. As of mid-1996, a total of 35 approvals had been granted to commercially grow 8 transgenic crops and one flower crop (carnation), with 8 different traits in 6 countries plus the EU. In 2000, Vitamin A-enriched golden rice was developed, though as of 2016 it was not yet in commercial production. In 2013 the leaders of the three research teams that first applied genetic engineering to crops, Robert Fraley, Marc Van Montagu and Mary-Dell Chilton were awarded the World Food Prize for improving the "quality, quantity or availability" of food in the world.

Methods

Genetically engineered crops have genes added or removed using genetic engineering techniques, originally including gene guns, electroporation, microinjection and agrobacterium. More recently, CRISPR and TALEN offered much more precise and convenient editing techniques.

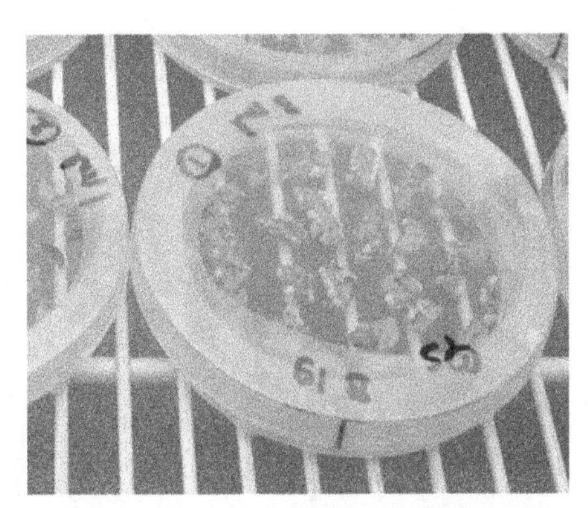

Plants (*Solanum chacoense*) being transformed using agrobacterium

Gene guns (also known as biolistics) "shoot" (direct high energy particles or radiations against) target genes into plant cells. It is the most common method. DNA is bound to tiny particles of gold or tungsten which are subsequently shot into plant tissue or single plant cells under high pressure. The accelerated particles penetrate both the cell wall and membranes. The DNA separates from the metal and is integrated into plant DNA inside the nucleus. This method has been applied successfully for many cultivated crops, especially monocots like wheat or maize, for which transformation using *Agrobacterium tumefaciens* has been less successful. The major disadvantage of this procedure is that serious damage can be done to the cellular tissue.

Agrobacterium tumefaciens-mediated transformation is another common technique. Agrobacteria are natural plant parasites, and their natural ability to transfer genes provides another engineering method. To create a suitable environment for themselves, these Agrobacteria insert their genes into plant hosts, resulting in a proliferation of modified plant cells near the soil level (crown gall). The genetic information for tumour growth is encoded on a mobile, circular DNA fragment (plasmid). When *Agrobacterium* infects a plant, it transfers this T-DNA to a random site in the plant genome. When used in genetic engineering the bacterial T-DNA is removed from the bacterial plasmid and replaced with the desired foreign gene. The bacterium is a vector, enabling transportation of foreign genes into plants. This method works especially well for dicotyledonous plants like potatoes, tomatoes, and tobacco. Agrobacteria infection is less successful in crops like wheat and maize.

Electroporation is used when the plant tissue does not contain cell walls. In this technique, "DNA enters the plant cells through miniature pores which are temporarily caused by electric pulses."

Microinjection directly injects the gene into the DNA.

Plant scientists, backed by results of modern comprehensive profiling of crop composition, point out that crops modified using GM techniques are less likely to have unintended changes than are conventionally bred crops.

In research tobacco and *Arabidopsis thaliana* are the most frequently modified plants, due to well-developed transformation methods, easy propagation and well studied genomes. They serve as model organisms for other plant species.

Introducing new genes into plants requires a promoter specific to the area where the gene is to be expressed. For instance, to express a gene only in rice grains and not in leaves, an endosperm-specific promoter is used. The codons of the gene must be optimized for the organism due to codon usage bias.

Types of Modifications

Transgenic maize containing a gene from the bacteria *Bacillus thuringiensis*

Transgenic

Transgenic plants have genes inserted into them that are derived from another species. The inserted genes can come from species within the same kingdom (plant to plant) or between kingdoms (for example, bacteria to plant). In many cases the inserted DNA has to be modified slightly in order to correctly and efficiently express in the host organism. Transgenic plants are used to express proteins like the cry toxins from *B. thuringiensis*, herbicide resistant genes, antibodies and antigens for vaccinations A study led by the European Food Safety Authority (EFSA) found also viral genes in transgenic plants.

Transgenic carrots have been used to produce the drug Taliglucerase alfa which is used to treat Gaucher's disease. In the laboratory, transgenic plants have been modified to increase photosynthesis (currently about 2% at most plants versus the theoretic potential of 9–10%). This is possible by changing the rubisco enzyme (i.e. changing C3 plants into C4 plants), by placing the rubisco in a carboxysome, by adding CO_2 pumps in the cell wall, by changing the leaf form/size. Plants have been engineered to exhibit bioluminescence that may become a sustainable alternative to electric lighting.

Cisgenic

Cisgenic plants are made using genes found within the same species or a closely related one, where conventional plant breeding can occur. Some breeders and scientists argue that cisgenic modification is useful for plants that are difficult to crossbreed by conventional means (such as potatoes), and that plants in the cisgenic category should not require the same regulatory scrutiny as transgenics.

Subgenic

Genetically modified plants can also be developed using gene knockdown or gene knockout to alter the genetic makeup of a plant without incorporating genes from other plants. In 2014, Chinese researcher Gao Caixia filed patents on the creation of a strain of wheat that is resistant to powdery mildew. The strain lacks genes that encode proteins that repress defenses against the mildew. The researchers deleted all three copies of the genes from wheat's hexaploid genome. Gao used the TALENs and CRISPR gene editing tools without adding or changing any other genes. No field trials were immediately planned. The CRISPR technique has also been used to modify white button mushrooms (*Agaricus bisporus*).

Economics

GM food's economic value to farmers is one of its major benefits, including in developing nations. A 2010 study found that Bt corn provided economic benefits of $6.9 billion over the previous 14 years in five Midwestern states. The majority ($4.3 billion) accrued to farmers producing non-Bt corn. This was attributed to European corn borer populations reduced by exposure to Bt corn, leaving fewer to attack conventional corn nearby. Agriculture economists calculated that "world surplus [increased by] $240.3 million for 1996. Of this total, the largest share (59%) went to U.S. farmers. Seed company Monsanto received the next largest share (21%), followed by US consumers (9%), the rest of the world (6%), and the germplasm supplier, Delta & Pine Land Company of Mississippi (5%)."

According to the International Service for the Acquisition of Agri-biotech Applications (ISAAA), in 2014 approximately 18 million farmers grew biotech crops in 28 countries; about 94% of the farmers were resource-poor in developing countries. 53% of the global biotech crop area of 181.5 million hectares was grown in 20 developing countries. PG Economics comprehensive 2012 study concluded that GM crops increased farm incomes worldwide by $14 billion in 2010, with over half this total going to farmers in developing countries.

Critics challenged the claimed benefits to farmers over the prevalence of biased observers and by the absence of randomized controlled trials. The main Bt crop grown by small farmers in developing countries is cotton. A 2006 review of Bt cotton findings by agricultural economists concluded, "the overall balance sheet, though promising, is mixed. Economic returns are highly variable over years, farm type, and geographical location".

In 2013 the European Academies Science Advisory Council (EASAC) asked the EU to allow the development of agricultural GM technologies to enable more sustainable agriculture, by employing fewer land, water and nutrient resources. EASAC also criticizes the EU's "timeconsuming and expensive regulatory framework" and said that the EU had fallen behind in the adoption of GM technologies.

Participants in agriculture business markets include seed companies, agrochemical companies, distributors, farmers, grain elevators and universities that develop new crops/traits and whose agricultural extensions advise farmers on best practices. According to a 2012 review based on data from the late 1990s and early 2000s, much of the GM crop grown each year is used for livestock feed and increased demand for meat leads to increased demand for GM feedcrops. Feed grain usage as a percentage of total crop production is 70% for corn and more than 90% of oil seed meals such as soybeans. About 65 million metric tons of GM corn grains and about 70 million metric tons of soybean meals derived from GM soybean become feed.

In 2014 the global value of biotech seed was US$15.7 billion; US$11.3 billion (72%) was in industrial countries and US$4.4 billion (28%) was in the developing countries. In 2009, Monsanto had $7.3 billion in sales of seeds and from licensing its technology; DuPont, through its Pioneer subsidiary, was the next biggest company in that market. As of 2009, the overall Roundup line of products including the GM seeds represented about 50% of Monsanto's business.

Some patents on GM traits have expired, allowing the legal development of generic strains that include these traits. For example, generic glyphosate-tolerant GM soybean is now available. Another impact is that traits developed by one vendor can be added to another vendor's proprietary strains, potentially increasing product choice and competition. The patent on the first type of *Roundup Ready* crop that Monsanto produced (soybeans) expired in 2014 and the first harvest of off-patent soybeans occurs in the spring of 2015. Monsanto has broadly licensed the patent to other seed companies that include the glyphosate resistance trait in their seed products. About 150 companies have licensed the technology, including Syngenta and DuPont Pioneer.

Yield

In 2014, the largest review yet concluded that GM crops' effects on farming were positive. The meta-analysis considered all published English-language examinations of the agronomic and economic impacts between 1995 and March 2014 for three major GM crops: soybean, maize, and cotton. The study found that herbicide-tolerant crops have lower production costs, while for insect-resistant crops the reduced pesticide use was offset by higher seed prices, leaving overall production costs about the same.

Yields increased 9% for herbicide tolerance and 25% for insect resistant varieties. Farmers who adopted GM crops made 69% higher profits than those who did not. The review found that GM crops help farmers in developing countries, increasing yields by 14 percentage points.

The researchers considered some studies that were not peer-reviewed, and a few that did not report sample sizes. They attempted to correct for publication bias, by considering sources beyond academic journals. The large data set allowed the study to control for potentially confounding variables such as fertiliser use. Separately, they concluded that the funding source did not influence study results.

Traits

GM crops grown today, or under development, have been modified with various traits. These traits include improved shelf life, disease resistance, stress resistance, herbicide resistance, pest resistance, production of useful goods such as biofuel or drugs, and ability to absorb toxins and for use in bioremediation of pollution.

Recently, research and development has been targeted to enhancement of crops that are locally important in developing countries, such as insect-resistant cowpea for Africa and insect-resistant brinjal (eggplant).

Lifetime

The first genetically modified crop approved for sale in the U.S. was the *FlavrSavr* tomato, which had a longer shelf life. It is no longer on the market.

In November 2014, the USDA approved a GM potato that prevents bruising.

In February 2015 Arctic Apples were approved by the USDA, becoming the first genetically modified apple approved for US sale. Gene silencing was used to reduce the expression of polyphenol oxidase (PPO), thus preventing enzymatic browning of the fruit after it has been sliced open. The trait was added to Granny Smith and Golden Delicious varieties. The trait includes a bacterial antibiotic resistance gene that provides resistance to the antibiotic kanamycin. The genetic engineering involved cultivation in the presence of kanamycin, which allowed only resistant cultivars to survive. Humans consuming apples do not acquire kanamycin resistance, per arcticapple.com. The FDA approved the apples in March 2015.

Nutrition

Edible Oils

Some GM soybeans offer improved oil profiles for processing or healthier eating. Camelina sativa has been modified to produce plants that accumulate high levels of oils similar to fish oils.

Vitamin Enrichment

Golden rice, developed by the International Rice Research Institute (IRRI), provides greater amounts of Vitamin A targeted at reducing Vitamin A deficiency. As of January 2016, golden rice has not yet been grown commercially in any country.

Researchers vitamin-enriched corn derived from South African white corn variety M37W, producing a 169-fold increase in Vitamin A, 6-fold increase in Vitamin C and doubled concentrations of folate. Modified Cavendish bananas express 10-fold the amount of Vitamin A as unmodified varieties.

Toxin Reduction

A genetically modified cassava under development offers lower cyanogen glucosides and enhanced protein and other nutrients (called BioCassava).

In November 2014, the USDA approved a potato, developed by J.R. Simplot Company, that prevents bruising and produces less acrylamide when fried. The modifications prevent natural, harmful proteins from being made via RNA interference. They do not employ genes from non-potato species. The trait was added to the Russet Burbank, Ranger Russet and Atlantic varieties.

Stress Resistance

Plants engineered to tolerate non-biological stressors such as drought, frost, high soil salinity, and nitrogen starvation were in development. In 2011, Monsanto's DroughtGard maize became the first drought-resistant GM crop to receive US marketing approval.

Herbicides

Glyphosate

As of 1999 the most prevalent GM trait was glyphosate-resistance. Glyphosate, (the active ingredient in Roundup and other herbicide products) kills plants by interfering with the shikimate pathway in plants, which is essential for the synthesis of the aromatic amino acids phenylalanine, tyrosine and tryptophan. The shikimate pathway is not present in animals, which instead obtain aromatic amino acids from their diet. More specifically, glyphosate inhibits the enzyme 5-enolpyruvylshikimate-3-phosphate synthase (EPSPS).

This trait was developed because the herbicides used on grain and grass crops at the time were highly toxic and not effective against narrow-leaved weeds. Thus, developing crops that could withstand spraying with glyphosate would both reduce environmental and health risks, and give an agricultural edge to the farmer.

Some micro-organisms have a version of EPSPS that is resistant to glyphosate inhibition. One of these was isolated from an *Agrobacterium* strain CP4 (CP4 EPSPS) that was resistant to glyphosate. The CP4 EPSPS gene was engineered for plant expression by fusing the 5' end of the gene to a chloroplast transit peptide derived from the petunia EPSPS. This transit peptide was used because it had shown previously an ability to deliver bacterial EPSPS to the chloroplasts of other plants. This CP4 EPSPS gene was cloned and transfected into soybeans.

The plasmid used to move the gene into soybeans was PV-GMGTO4. It contained three bacterial genes, two CP4 EPSPS genes, and a gene encoding beta-glucuronidase (GUS) from *Escherichia coli* as a marker. The DNA was injected into the soybeans using the particle acceleration method. Soybean cultivar A5403 was used for the transformation.

Bromoxynil

Tobacco plants have been engineered to be resistant to the herbicide bromoxynil.

Glufosinate

Crops have been commercialized that are resistant to the herbicide glufosinate, as well. Crops engineered for resistance to multiple herbicides to allow farmers to use a mixed group of two, three, or four different chemicals are under development to combat growing herbicide resistance.

2,4-D

In October 2014 the US EPA registered Dow's Enlist Duo maize, which is genetically modified to be resistant to both glyphosate and 2,4-D, in six states. Inserting a bacterial aryloxyalkanoate dioxygenase gene, *aad1* makes the corn resistant to 2,4-D. The USDA had approved maize and soybeans with the mutation in September 2014.

Dicamba

Monsanto has requested approval for a stacked strain that is tolerant of both glyphosate and dicamba.

Pest Resistance

Insects

Tobacco, corn, rice and many other crops have been engineered to express genes encoding for insecticidal proteins from Bacillus thuringiensis (Bt). Papaya, potatoes, and squash have been engineered to resist viral pathogens such as cucumber mosaic virus which, despite its name, infects a wide variety of plants. The introduction of Bt crops during the period between 1996 and 2005 has been estimated to have reduced the total volume of insecticide active ingredient use in the United States by over 100 thousand tons. This represents a 19.4% reduction in insecticide use.

In the late 1990s, a genetically modified potato that was resistant to the Colorado potato beetle was withdrawn because major buyers rejected it, fearing consumer opposition.

Viruses

Virus resistant papaya were developed In response to a papaya ringspot virus (PRV) outbreak in Hawaii in the late 1990s. . They incorporate PRV DNA. By 2010, 80% of Hawaiian papaya plants were genetically modified.

Potatoes were engineered for resistance to potato leaf roll virus and Potato virus Y in 1998. Poor sales led to their market withdrawal after three years.

Yellow squash that were resistant to at first two, then three viruses were developed, beginning in the 1990s. The viruses are watermelon, cucumber and zucchini/courgette yellow mosaic. Squash was the second GM crop to be approved by US regulators. The trait was later added to zucchini.

Many strains of corn have been developed in recent years to combat the spread of Maize dwarf mosaic virus, a costly virus that causes stunted growth which is carried in Johnson grass and spread by aphid insect vectors. These strands are commercially available although the resistance is not standard among GM corn variants.

By-products

Drugs

In 2012, the FDA approved the first plant-produced pharmaceutical, a treatment for Gaucher's Disease. Tobacco plants have been modified to produce therapeutic antibodies.

Biofuel

Algae is under development for use in biofuels. Researchers in Singapore were working on GM jatropha for biofuel production. Syngenta has USDA approval to market a maize trademarked Enogen that has been genetically modified to convert its starch to sugar for ethanol. In 2013, the Flemish Institute for Biotechnology was investigating poplar trees genetically engineered to contain less lignin to ease conversion into ethanol. Lignin is the critical limiting factor when using wood to make bio-ethanol because lignin limits the accessibility of cellulose microfibrils to depolymerization by enzymes.

Materials

Companies and labs are working on plants that can be used to make bioplastics. Potatoes that produce industrially useful starches have been developed as well. Oilseed can be modified to produce fatty acids for detergents, substitute fuels and petrochemicals.

Bioremediation

Scientists at the University of York developed a weed (*Arabidopsis thaliana*) that contains genes from bacteria that could clean TNT and RDX-explosive soil contaminants in 2011. 16 million hectares in the USA (1.5% of the total surface) are estimated to be contaminated with TNT and RDX. However *A. thaliana* was not tough enough for use on military test grounds. Modifications in 2016 included switchgrass and bentgrass.

Genetically modified plants have been used for bioremediation of contaminated soils. Mercury, selenium and organic pollutants such as polychlorinated biphenyls (PCBs).

Marine environments are especially vulnerable since pollution such as oil spills are not containable. In addition to anthropogenic pollution, millions of tons of petroleum annually enter the marine environment from natural seepages. Despite its toxicity, a considerable fraction of petroleum oil entering marine systems is eliminated by the hydrocarbon-degrading activities of microbial communities. Particularly successful is a recently discovered group of specialists, the so-called hydrocarbonoclastic bacteria (HCCB) that may offer useful genes.

Asexual Reproduction

Crops such as maize reproduce sexually each year. This randomizes which genes get propagated to the next generation, meaning that desirable traits can be lost. To maintain a high-quality crop, some farmers purchase seeds every year. Typically, the seed company maintains two inbred varieties, and crosses them into a hybrid strain that is then sold. Related plants like sorghum and gamma grass are able to perform apomixis, a form of asexual reproduction that keeps the plant's DNA intact. This trait is apparently controlled by a single dominant gene, but traditional breeding has been unsuccessful in creating asexually-reproducing maize. Genetic engineering offers another route to this goal. Successful modification would allow farmers to replant harvested seeds that retain desirable traits, rather than relying on purchased seed.

Crops

Herbicide Tolerance

GMO	Use	Countries approved in	First approved	Notes
Alfalfa	Animal feed	USA	2005	Approval withdrawn in 2007 and then re-approved in 2011
Canola	Cooking oil Margarine Emulsifiers in packaged foods	Australia	2003	
		Canada	1995	
		USA	1995	

Cotton	Fiber Cottonseed oil Animal feed	Argentina	2001	
		Australia	2002	
		Brazil	2008	
		Columbia	2004	
		Costa Rica	2008	
		Mexico	2000	
		Paraguay	2013	
		South Africa	2000	
		USA	1994	
Maize	Animal feed high-fructose corn syrup corn starch	Argentina	1998	
		Brazil	2007	
		Canada	1996	
		Colombia	2007	
		Cuba	2011	
		European Union	1998	Grown in Portugal, Spain, Czech Republic, Slovakia and Romania
		Honduras	2001	
		Paraguay	2012	
		Philippines	2002	
		South Africa	2002	
		USA	1995	
		Uruguay	2003	
Soybean	Animal feed Soybean oil	Argentina	1996	
		Bolivia	2005	
		Brazil	1998	
		Canada	1995	
		Chile	2007	
		Costa Rica	2001	
		Mexico	1996	
		Paraguay	2004	
		South Africa	2001	
		USA	1993	
		Uruguay	1996	
Sugar Beet	Food	Canada	2001	
		USA	1998	Commercialised 2007, production blocked 2010, resumed 2011.

Insect Resistance

GMO	Use	Countries approved in	First approved	Notes
Cotton	Fiber Cottonseed oil Animal feed	Argentina	1998	
		Australia	2003	
		Brazil	2005	
		Burkina Faso	2009	
		China	1997	
		Colombia	2003	
		Costa Rica	2008	
		India	2002	Largest producer of Bt cotton
		Mexico	1996	
		Myanmar	2006	
		Pakistan	2010	
		Paraguay	2007	
		South Africa	1997	
		Sudan	2012	
		USA	1995	
Eggplant	Food	Bangladesh	2013	12 ha planted on 120 farms in 2014
Maize	Animal feed high-fructose corn syrup corn starch	Argentina	1998	
		Brazil	2005	
		Columbia	2003	
		Mexico	1996	Centre of origin for maize
		Paraguay	2007	
		Philippines	2002	
		South Africa	1997	
		Uruguay	2003	
		USA	1995	
Poplar	Tree	China	1998	543 ha of bt poplar planted in 2014

Other Modified Traits

GMO	Use	Trait	Countries approved in	First approved	Notes
Canola	Cooking oil Margarine Emulsifiers in packaged foods	High laurate canola	Canada	1996	
			USA	1994	
		Phytase production	USA	1998	
Carnation	Ornamental	Delayed senescence	Australia	1995	
			Norway	1998	
		Modified flower colour	Australia	1995	
			Columbia	2000	In 2014 4 ha were grown in greenhouses for export
			European Union	1998	Two events expired 2008, another approved 2007
			Japan	2004	
			Malaysia	2012	For ornamental purposes
			Norway	1997	
Maize	Animal feed high-fructose corn syrup corn starch	Increased lysine	Canada	2006	
			USA	2006	
		Drought tolerance	Canada	2010	
			USA	2011	
Papaya	Food	Virus resistance	China	2006	
			USA	1996	Mostly grown in Hawaii
Petunia	Ornamental	Modified flower colour	China	1997	
Potato	Food	Virus resistance	Canada	1999	
			USA	1997	
	Industrial	Modified starch	USA	2014	
Rose	Ornamental	Modified flower colour	Australia	2009	Surrendered renewal
			Colombia	2010	Greenhouse cultivation for export only.
			Japan	2008	
			USA	2011	
Soybean	Animal feed Soybean oil	Increased oleic acid production	Argentina	2015	
			Canada	2000	
			USA	1997	
		Stearidonic acid production	Canada	2011	
			USA	2011	
Squash	Food	Virus resistance	USA	1994	

Sugar Cane	Food	Drought tolerance	Indonesia	2013	Environmental certificate only
Tobacco	Cigarettes	Nicotine reduction	USA	2002	

Development

The number of USDA-approved field releases for testing grew from 4 in 1985 to 1,194 in 2002 and averaged around 800 per year thereafter. The number of sites per release and the number of gene constructs (ways that the gene of interest is packaged together with other elements)—have rapidly increased since 2005. Releases with agronomic properties (such as drought resistance) jumped from 1,043 in 2005 to 5,190 in 2013. As of September 2013, about 7,800 releases had been approved for corn, more than 2,200 for soybeans, more than 1,100 for cotton, and about 900 for potatoes. Releases were approved for herbicide tolerance (6,772 releases), insect resistance (4,809), product quality such as flavor or nutrition (4,896), agronomic properties like drought resistance (5,190), and virus/fungal resistance (2,616). The institutions with the most authorized field releases include Monsanto with 6,782, Pioneer/DuPont with 1,405, Syngenta with 565, and USDA's Agricultural Research Service with 370. As of September 2013 USDA had received proposals for releasing GM rice, squash, plum, rose, tobacco, flax and chicory.

Farming Practices

Bt Resistance

Constant exposure to a toxin creates evolutionary pressure for pests resistant to that toxin. Overreliance on glyphosate and a reduction in the diversity of weed management practices allowed the spread of glyphosate resistance in 14 weed species/biotypes in the US.

One method of reducing resistance is the creation of refuges to allow nonresistant organisms to survive and maintain a susceptible population.

To reduce resistance to Bt crops, the 1996 commercialization of transgenic cotton and maize came with a management strategy to prevent insects from becoming resistant. Insect resistance management plans are mandatory for Bt crops. The aim is to encourage a large population of pests so that any (recessive) resistance genes are diluted within the population. Resistance lowers evolutionary fitness in the absence of the stressor (Bt). In refuges, non-resistant strains outcompete resistant ones.

With sufficiently high levels of transgene expression, nearly all of the heterozygotes (S/s), i.e., the largest segment of the pest population carrying a resistance allele, will be killed before maturation, thus preventing transmission of the resistance gene to their progeny. Refuges (i. e., fields of nontransgenic plants) adjacent to transgenic fields increases the likelihood that homozygous resistant (s/s) individuals and any surviving heterozygotes will mate with susceptible (S/S) individuals from the refuge, instead of with other individuals carrying the resistance allele. As a result, the resistance gene frequency in the population remains lower.

Complicating factors can affect the success of the high-dose/refuge strategy. For example, if the temperature is not ideal, thermal stress can lower Bt toxin production and leave the plant more susceptible. More importantly, reduced late-season expression has been documented, possibly resulting from DNA methylation of the promoter. The success of the high-dose/refuge strategy has successfully maintained the value of Bt crops. This success has depended on factors independent of management strategy, including low initial resistance allele frequencies, fitness costs associated with resistance, and the abundance of non-Bt host plants outside the refuges.

Companies that produce Bt seed are introducing strains with multiple Bt proteins. Monsanto did this with Bt cotton in India, where the product was rapidly adopted. Monsanto has also; in an attempt to simplify the process of implementing refuges in fields to comply with Insect Resistance Management(IRM) policies and prevent irresponsible planting practices; begun marketing seed bags with a set proportion of refuge (non-transgenic) seeds mixed in with the Bt seeds being sold. Coined "Refuge-In-a-Bag" (RIB), this practice is intended to increase farmer compliance with refuge requirements and reduce additional labor needed at planting from having separate Bt and refuge seed bags on hand. This strategy is likely to reduce the likelihood of Bt-resistance occurring for corn rootworm, but may increase the risk of resistance for lepidopteran corn pests, such as European corn borer. Increased concerns for resistance with seed mixtures include partially resistant larvae on a Bt plant being able to move to a susceptible plant to survive or cross pollination of refuge pollen on to Bt plants that can lower the amount of Bt expressed in kernels for ear feeding insects.

Herbicide Resistance

Best management practices (BMPs) to control weeds may help delay resistance. BMPs include applying multiple herbicides with different modes of action, rotating crops, planting weed-free seed, scouting fields routinely, cleaning equipment to reduce the transmission of weeds to other fields, and maintaining field borders. The most widely planted GMOs are designed to tolerate herbicides. By 2006 some weed populations had evolved to tolerate some of the same herbicides. Palmer amaranth is a weed that competes with cotton. A native of the southwestern US, it traveled east and was first found resistant to glyphosate in 2006, less than 10 years after GM cotton was introduced.

Plant Protection

Farmers generally use less insecticide when they plant Bt-resistant crops. Insecticide use on corn farms declined from 0.21 pound per planted acre in 1995 to 0.02 pound in 2010. This is consistent with the decline in European corn borer populations as a direct result of Bt corn and cotton. The establishment of minimum refuge requirements helped delay the evolution of Bt resistance. However resistance appears to be developing to some Bt traits in some areas.

Tillage

By leaving at least 30% of crop residue on the soil surface from harvest through planting, conservation tillage reduces soil erosion from wind and water, increases water retention, and reduces soil degradation as well as water and chemical runoff. In addition, conservation tillage reduces the carbon footprint of agriculture. A 2014 review covering 12 states from 1996 to 2006, found that a 1% increase in herbicde-tolerant (HT) soybean adoption leads to a 0.21% increase in conservation tillage and a 0.3% decrease in quality-adjusted herbicide use.

Regulation

The regulation of genetic engineering concerns the approaches taken by governments to assess and manage the risks associated with the development and release of genetically modified crops. There are differences in the regulation of GM crops between countries, with some of the most marked differences occurring between the USA and Europe. Regulation varies in a given country depending on the intended use of each product. For example, a crop not intended for food use is generally not reviewed by authorities responsible for food safety.

Production

In 2013, GM crops were planted in 27 countries; 19 were developing countries and 8 were developed countries. 2013 was the second year in which developing countries grew a majority (54%) of the total GM harvest. 18 million farmers grew GM crops; around 90% were small-holding farmers in developing countries.

Country	2013– GM planted area (million hectares)	Biotech crops
USA	70.1	Maize, Soybean, Cotton, Canola, Sugarbeet, Alfalfa, Papaya, Squash
Brazil	40.3	Soybean, Maize, Cotton
Argentina	24.4	Soybean, Maize, Cotton
India	11.0	Cotton
Canada	10.8	Canola, Maize, Soybean, Sugarbeet
Total	175.2	----

The United States Department of Agriculture (USDA) reports every year on the total area of GMO varieties planted in the United States. According to National Agricultural Statistics Service, the states published in these tables represent 81–86 percent of all corn planted area, 88–90 percent of all soybean planted area, and 81–93 percent of all upland cotton planted area (depending on the year).

Global estimates are produced by the International Service for the Acquisition of Agri-biotech Applications (ISAAA) and can be found in their annual reports, "Global Status of Commercialized Transgenic Crops".

Farmers have widely adopted GM technology. Between 1996 and 2013, the total surface area of land cultivated with GM crops increased by a factor of 100, from 17,000 square kilometers (4,200,000 acres) to 1,750,000 km^2 (432 million acres). 10% of the world's arable land was planted with GM crops in 2010. As of 2011, 11 different transgenic crops were grown commercially on 395 million acres (160 million hectares) in 29 countries such as the USA, Brazil, Argentina, India, Canada, China, Paraguay, Pakistan, South Africa, Uruguay, Bolivia, Australia, Philippines, Myanmar, Burkina Faso, Mexico and Spain. One of the key reasons for this widespread adoption is the perceived economic benefit the technology brings to farmers. For example, the system of planting glyphosate-resistant seed and then applying glyphosate once plants emerged provided farmers with the opportunity to dramatically increase the yield from a given plot of land, since this allowed them to plant rows closer together. Without it, farmers had to plant rows far enough apart to control post-emergent weeds with mechanical tillage. Likewise, using Bt seeds means that farmers do not have to purchase insecticides, and then invest time, fuel, and equipment in applying them. However critics have disputed whether yields are higher and whether chemical use is less, with GM crops.

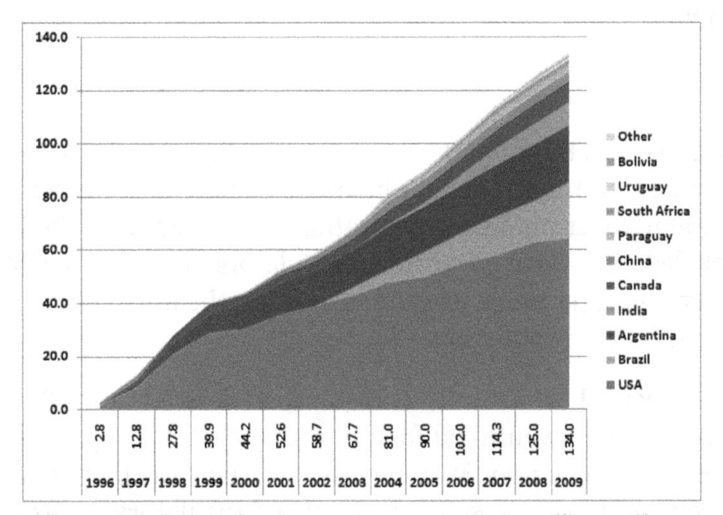

Land area used for genetically modified crops by country (1996–2009), in millions of hectares. In 2011, the land area used was 160 million hectares, or 1.6 million square kilometers.

In the US, by 2014, 94% of the planted area of soybeans, 96% of cotton and 93% of corn were genetically modified varieties. Genetically modified soybeans carried herbicide-tolerant traits only, but maize and cotton carried both herbicide tolerance and insect protection traits (the latter largely Bt protein). These constitute "input-traits" that are aimed to financially benefit the producers, but may have indirect environmental benefits and cost benefits to consumers. The Grocery Manufacturers of America estimated in 2003 that 70–75% of all processed foods in the U.S. contained a GM ingredient.

Europe grows relatively few genetically engineered crops with the exception of Spain, where one fifth of maize is genetically engineered, and smaller amounts in five other countries. The EU had a 'de facto' ban on the approval of new GM crops, from 1999 until 2004. GM crops are now regulated by the EU. In 2015, genetically engineered crops are banned in 38 countries worldwide, 19 of them in Europe. Developing countries grew 54 percent of genetically engineered crops in 2013.

In recent years GM crops expanded rapidly in developing countries. In 2013 approximately 18 million farmers grew 54% of worldwide GM crops in developing countries. 2013's largest increase was in Brazil (403,000 km² versus 368,000 km² in 2012). GM cotton began growing in India in 2002, reaching 110,000 km² in 2013.

According to the 2013 ISAAA brief: "...a total of 36 countries (35 + EU-28) have granted regulatory approvals for biotech crops for food and/or feed use and for environmental release or planting since 1994... a total of 2,833 regulatory approvals involving 27 GM crops and 336 GM events (NB: an "event" is a specific genetic modification in a specific species) have been issued by authorities, of which 1,321 are for food use (direct use or processing), 918 for feed use (direct use or processing) and 599 for environmental release or planting. Japan has the largest number (198), followed by the U.S.A. (165, not including "stacked" events), Canada (146), Mexico (131), South Korea (103), Australia (93), New Zealand (83), European Union (71 including approvals that have expired or under renewal process), Philippines (68), Taiwan (65), Colombia (59), China (55) and South Africa (52). Maize has the largest number (130 events in 27 countries), followed by cotton (49 events in 22 countries), potato (31 events in 10 countries), canola (30 events in 12 countries) and soybean (27 events in 26 countries).

Controversy

GM foods are controversial and the subject of protests, vandalism, referenda, legislation, court action and scientific disputes. The controversies involve consumers, biotechnology companies, governmental regulators, non-governmental organizations and scientists. The key areas are whether GM food should be labeled, the role of government regulators, the effect of GM crops on health and the environment, the effects of pesticide use and resistance, the impact on farmers, and their roles in feeding the world and energy production.

There is a scientific consensus that currently available food derived from GM crops poses no greater risk to human health than conventional food, but that each GM food needs to be tested on a case-by-case basis before introduction. Nonetheless, members of the public are much less likely than scientists to perceive GM foods as safe. The legal and regulatory status of GM foods varies by country, with some nations banning or restricting them, and others permitting them with widely differing degrees of regulation.

No reports of ill effects have been documented in the human population from GM food. Although GMO labeling is required in many countries, the United States Food and Drug Administration does not require labeling, nor does it recognize a distinction between approved GMO and non-GMO foods.

Advocacy groups such as Center for Food Safety, Union of Concerned Scientists, Greenpeace and the World Wildlife Fund claim that risks related to GM food have not been adequately examined and managed, that GMOs are not sufficiently tested and should be labelled, and that regulatory authorities and scientific bodies are too closely tied to industry. Some studies have claimed that genetically modified crops can cause harm; a 2016 review that reanalyzed the data from six of these studies found that their statistical methodologies were flawed and did not demonstrate harm, and said that conclusions about GMO crop safety should be drawn from "the totality of the evidence... instead of far-fetched evidence from single studies".

Genetically Modified Soybean

A genetically modified soybean is a soybean (*Glycine max*) that has had DNA introduced into it using genetic engineering techniques. In 1994 the first genetically modified soybean was introduced to the U.S. market, by Monsanto. In 2014, 90.7 million hectares of GM soy were planted worldwide, 82% of the total soy cultivation area.

Examples of Transgenic Soybeans

The genetic makeup of a soybean gives it a wide variety of uses, thus keeping it in high demand. First, manufacturers only wanted to use transgenics to be able to grow more soy at a minimal cost to meet this demand, and to fix any problems in the growing process, but they eventually found they could modify the soybean to contain healthier components, or even focus on one aspect of the soybean to produce in larger quantities. These phases became known as the first and second generation of genetically modified (GM) foods. As Peter Celec describes, "benefits of the first generation of GM foods were oriented towards the production process and companies, the second generation of GM foods offers, on contrary, various advantages and added value for the consumer", including "improved nutritional composition or even therapeutic effects."

Roundup Ready Soybean

Roundup Ready Soybeans (The first variety was also known as GTS 40-3-2 (OECD UI: MON-04032-6)) are a series of genetically engineered varieties of glyphosate-resistant soybeans produced by Monsanto.

Glyphosate kills plants by interfering with the synthesis of the essential amino acids phenylalanine, tyrosine and tryptophan. These amino acids are called "essential" because animals cannot make them; only plants and micro-organisms can make them and animals obtain them by eating plants.

Plants and microorganisms make these amino acids with an enzyme that only plants and lower organisms have, called 5-enolpyruvylshikimate-3-phosphate synthase (EPSPS). EPSPS is not present in animals, which instead obtain aromatic amino acids from their diet.

Roundup Ready Soybeans express a version of EPSPS from the CP4 strain of the bacteria, *Agrobacterium tumefaciens*, expression of which is regulated by an enhanced 35S promoter (E35S) from cauliflower mosaic virus (CaMV), a chloroplast transit peptide (CTP4) coding sequence from Petunia hybrida, and a nopaline synthase (nos 3') transcriptional termination element from Agrobacterium tumefaciens. The plasmid with EPSPS and the other genetic elements mentioned above was inserted into soybean germplasm with a gene gun by scientists at Monsanto and Asgrow. The patent on the first generation of Roundup Ready soybeans expired in March 2015.

History

First approved commercially in the United States during 1994, GTS 40-3-2 was subsequently introduced to Canada in 1995, Japan and Argentina in 1996, Uruguay in 1997, Mexico and Brazil in 1998, and South Africa in 2001.

Detection

GTS 40-3-2 can be detected using both nucleic acid and protein analysis methods.

Generic GMO Soybeans

Following expiration of Monsanto's patent on the first variety of glyphosate-resistant Roundup Ready soybeans, development began on glyphosate-resistant "generic" soybeans. The first variety, developed at the University of Arkansas Division of Agriculture, came on the market in 2015. With a slightly lower yield than newer Monsanto varieties, it costs about half as much, and seeds can be saved for subsequent years. According to its creator it is adapted to conditions in Arkansas. Several other varieties are being bred by crossing the original variety of Roundup Ready soybeans with other soybean varieties.

Stacked Traits

Monsanto developed a glyphosate-resistant soybean that also expresses Cry1Ac protein from Bacillus thuringiensis and the glyphosate-resistance gene, which completed the Brazilian regulatory process in 2010.

Genetic Modification to Improve Soybean Oil

Soy has been genetically modified to improve the quality of soy oil. Soy oil has a fatty acid profile that makes it susceptible to oxidation, which makes it rancid, and this has limited its usefulness to the food industry. Genetic modifications increased the amount of oleic acid and stearic acid and decreased the amount of linolenic acid. By silencing, or knocking out, the delta 9 and delta 12 desaturases. DuPont Pioneer created a high oleic fatty acid soybean with levels of oleic acid greater than 80%, and started marketing it in 2010.

Regulation

The regulation of genetic engineering concerns the approaches taken by governments to assess and manage the risks associated with the development and release of genetically modified crops. There are differences in the regulation of GM crops between countries, with some of the most marked differences occurring between the USA and Europe. Soy beans are allowed a Maximum Residue Limit of glyphosate of 20 mg/Kg for international trade. Regulation varies in a given country depending on the intended use of the products of the genetic engineering. For example, a crop not intended for food use is generally not reviewed by authorities responsible for food safety.

Controversy

There is a scientific consensus that currently available food derived from GM crops poses no greater risk to human health than conventional food, but that each GM food needs to be tested on a case-by-case basis before introduction. Nonetheless, members of the public are much less likely than scientists to perceive GM foods as safe. The legal and regulatory status of GM foods varies by country, with some nations banning or restricting them, and others permitting them with widely differing degrees of regulation.

A 2010 study found that in the United States, GM crops also provide a number of ecological benefits.

Critics have objected to GM crops on several grounds, including ecological concerns, and economic concerns raised by the fact these organisms are subject to intellectual property law. GM crops also are involved in controversies over GM food with respect to whether food produced from GM crops is safe and whether GM crops are needed to address the world's food needs. See the genetically modified food controversies article for discussion of issues about GM crops and GM food. These controversies have led to litigation, international trade disputes, and protests, and to restrictive legislation in most countries.

Genetically Modified Maize

Transgenic maize containing a gene from the bacteria *Bacillus thuringiensis*

Genetically modified maize (corn) is a genetically modified crop. Specific maize strains have been genetically engineered to express agriculturally-desirable traits, including resistance to pests and to herbicides. Maize strains with both traits are now in use in multiple countries. GM maize has also caused controversy with respect to possible health effects, impact on other insects and impact on other plants via gene flow. One strain, called Starlink, was approved only for animal feed in the US, but was found in food, leading to a series of recalls starting in 2000.

Marketed Products

Herbicide Resistant Maize

Corn varieties resistant to glyphosate herbicides were first commercialized in 1996 by Monsanto, and are known as "Roundup Ready Corn". They tolerate the use of Roundup. Bayer CropScience developed "Liberty Link Corn" that is resistant to glufosinate. Pioneer Hi-Bred has developed and markets corn hybrids with tolerance to imidazoline herbicides under the trademark "Clearfield" – though in these hybrids, the herbicide-tolerance trait was bred using tissue culture selection and the chemical mutagen ethyl methanesulfonate, not genetic engineering. Consequently, the regulatory framework governing the approval of transgenic crops does not apply for Clearfield.

As of 2011, herbicide-resistant GM corn was grown in 14 countries. By 2012, 26 varieties herbicide-resistant GM maize were authorised for import into the European Union., but such imports remain controversial. Cultivation of herbicide-resistant corn in the EU provides substantial farm-level benefits.

Insecticide-producing Corn

Bt corn is a variant of maize that has been genetically altered to express one or more proteins from the bacterium *Bacillus thuringiensis*. The protein is poisonous to certain insect pests and is widely used in organic gardening. The European corn borer causes about a billion dollars in damage to corn crops each year.

The European corn borer, *Ostrinia nubilalis*, destroys corn crops by burrowing into the stem, causing the plant to fall over.

In recent years, traits have been added to ward off Corn ear worms and root worms, the latter of which annually causes about a billion dollars in damages.

The Bt protein is expressed throughout the plant. When a vulnerable insect eats the Bt-containing plant, the protein is activated in its gut, which is alkaline. In the alkaline environment the protein partially unfolds and is cut by other proteins, forming a toxin that paralyzes the insect's digestive system and forms holes in the gut wall. The insect stops eating within a few hours and eventually starves.

In 1996, the first GM maize producing a Bt Cry protein was approved, which killed the European corn borer and related species; subsequent Bt genes were introduced that killed corn rootworm larvae.

Approved Bt genes include single and stacked (event names bracketed) configurations of: Cry1A.105 (MON89034), CryIAb (MON810), CryIF (1507), Cry2Ab (MON89034), Cry3Bb1 (MON863 and MON88017), Cry34Ab1 (59122), Cry35Ab1 (59122), mCry3A (MIR604), and Vip3A (MIR162), in both corn and cotton. Corn genetically modified to produce VIP was first approved in the US in 2010.

Drought Resistance

In 2013 Monsanto launched the first transgenic drought tolerance trait in a line of corn hybrids called DroughtGard. The MON 87460 trait is provided by the insertion of the cspB gene from the soil microbe *Bacillus subtilis*; it was approved by the USDA in 2011 and by China in 2013.

Sweet Corn

GM sweet corn varieties include "Attribute", the brand name for insect-resistant sweet corn developed by Syngenta. and Performance Series™ insect-resistant sweet corn developed by Monsanto.

Products in Development

In 2007, South African researchers announced the production of transgenic maize resistant to maize streak virus (MSV), although it has not been released as a product.

While breeding cultivars for resistance to MSV isn't done in the public, the private sector, international research centers, and national programmes have done all of the breeding.

As of 2014, there have been a few MSV-tolerant cultivars released in Africa. A private company Seedco has released 5 MSV cultivars. While these seeds are more expensive, the price of maize is subsidized by the government so even the poorest of farmers can afford it.

Refuges

US Environmental Protection Agency (EPA) regulations require farmers who plant Bt corn to plant non-Bt corn nearby (called a refuge) to provide a location to harbor vulnerable pests. Typically, 20% of corn in a grower's fields must be refuge; refuge must be at least 0.5 miles from Bt corn for lepidopteran pests, and refuge for corn rootworm must at least be adjacent to a Bt field.

The theory behind these refuges is to slow the evolution of resistance to the pesticide. EPA regulations also require seed companies to train farmers how to maintain refuges, to collect data on the refuges and to report that data to the EPA. A study of these reports found that from 2003 to 2005 farmer compliance with keeping refuges was above 90%, but that by 2008 approximately 25% of Bt corn farmers did not keep refuges properly, raising concerns that resistance would develop.

Unmodified crops received most of the economic benefits of Bt corn in the US in 1996-2007, because of the overall reduction of pest populations. This reduction came because females laid eggs on modified and unmodified strains alike.

Seed bags containing both Bt and refuge seed have been approved by the EPA in the United States. These seed mixtures were marketed as "Refuge in a Bag" (RIB) to increase farmer compliance with refuge requirements and reduce additional work needed at planting from having separate Bt and refuge seed bags on hand. The EPA approved a lower percentage of refuge seed in these seed mixtures ranging from 5 to 10%. This strategy is likely to reduce the likelihood of Bt-resistance occurring for corn rootworm, but may increase the risk of resistance for lepidopteran pests, such as European corn borer. Increased concerns for resistance with seed mixtures include partially resistant larvae on a Bt plant being able to move to a susceptible plant to survive or cross pollination of refuge pollen on to Bt plants that can lower the amount of Bt expressed in kernels for ear feeding insects.

Resistance

Resistant strains of the European corn borer have developed in areas with defective or absent refuge management.

In November 2009, Monsanto scientists found the pink bollworm had become resistant to first-generation Bt cotton in parts of Gujarat, India – that generation expresses one Bt gene, *Cry1Ac*. This was the first instance of Bt resistance confirmed by Monsanto anywhere in the world. Bollworm resistance to first generation Bt cotton has been identified in the Australia, China, Spain and the United States. In 2012, a Florida field trial demonstrated that army worms were resistant to pesticide-containing GM corn produced by Dupont-Dow; armyworm resistance was first discovered in Puerto Rico in 2006, prompting Dow and DuPont to voluntarily stop selling the product on the island.

Regulation

Regulation of GM crops varies between countries, with some of the most-marked differences occurring between the USA and Europe. Regulation varies in a given country depending on intended uses.

Controversy

There is a scientific consensus that currently available food derived from GM crops poses no greater risk to human health than conventional food, but that each GM food needs to be tested on a case-by-case basis before introduction. Nonetheless, members of the public are much less likely than scientists to perceive GM foods as safe. The legal and regulatory status of GM foods varies by country, with some nations banning or restricting them, and others permitting them with widely differing degrees of regulation.

The scientific rigor of the studies regarding human health has been disputed due to alleged lack of independence and due to conflicts of interest involving governing bodies and some of those who perform and evaluate the studies.

GM crops provide a number of ecological benefits, but there are also concerns for their overuse, stalled research outside of the Bt seed industry, proper management and issues with Bt resistance arising from their misuse.

Critics have objected to GM crops on ecological, economic and health grounds. The economic issues derive from those organisms that are subject to intellectual property law, mostly patents. The first generation of GM crops lose patent protection beginning in 2015. Monsanto has claimed it will not pursue farmers who retain seeds of off-patent varieties. These controversies have led to litigation, international trade disputes, protests and to restrictive legislation in most countries.

Effects on Nontarget Insects

Critics claim that Bt proteins could target predatory and other beneficial or harmless insects as well as the targeted pest. These proteins have been used as organic sprays for insect control in France since 1938 and the USA since 1958 with no ill effects on the environment reported. While *cyt* proteins are toxic towards the insect orders Coleoptera (beetles) and Diptera (flies), *cry* proteins selectively target Lepidopterans (moths and butterflies). As a toxic mechanism, *cry* proteins bind to specific receptors on the membranes of mid-gut (epithelial) cells, resulting in rupture of those cells. Any organism that lacks the appropriate gut receptors cannot be affected by the *cry* protein, and therefore Bt. Regulatory agencies assess the potential for the transgenic plant to impact nontarget organisms before approving commercial release.

A 1999 study found that in a lab environment, pollen from Bt maize dusted onto milkweed could harm the monarch butterfly. Several groups later studied the phenomenon in both the field and the laboratory, resulting in a risk assessment that concluded that any risk posed by the corn to butterfly populations under real-world conditions was negligible. A 2002 review of the scientific literature concluded that "the commercial large-scale cultivation of current Bt–maize hybrids did not pose a significant risk to the monarch population". A 2007 review found that "nontarget invertebrates are generally more abundant in Bt cotton and Bt maize fields than in nontransgenic fields managed with insecticides. However, in comparison with insecticide-free control fields, certain nontarget taxa are less abundant in Bt fields."

Gene Flow

Gene flow is the transfer of genes and/or alleles from one species to another. Concerns focus on the interaction between GM and other maize varieties in Mexico, and of gene flow into refuges.

In 2009 the government of Mexico created a regulatory pathway for genetically modified maize, but because Mexico is the center of diversity for maize, gene flow could affect a large fraction of the world's maize strains. A 2001 report in *Nature* presented evidence that Bt maize was cross-breeding with unmodified maize in Mexico. The data in this paper was later described as originating from an artifact. *Nature* later stated, "the evidence available is not sufficient to justify the publication of the original paper". A 2005 large-scale study failed to find any evidence of contamination in Oaxaca. However, other authors also found evidence of cross-breeding between natural maize and transgenic maize.

A 2004 study found Bt protein in kernels of refuge corn.

Food

The French High Council of Biotechnologies Scientific Committee reviewed the 2009 Vendômois *et al.* study and concluded that it "..presents no admissible scientific element likely to ascribe any haematological, hepatic or renal toxicity to the three re-analysed GMOs." However, the French government applies the precautionary principle with respect to GMOs.

A review by Food Standards Australia New Zealand and others of the same study concluded that the results were due to chance alone.

A 2011 Canadian study looked at the presence of CryAb1 protein (BT toxin) in non-pregnant women, pregnant women and fetal blood. All groups had detectable levels of the protein, including 93% of pregnant women and 80% of fetuses at concentrations of 0.19 ± 0.30 and 0.04 ± 0.04 mean ± SD ng/ml, respectively. The paper did not discuss safety implications or find any health problems. The paper was found to be unconvincing by multiple authors and organizations. In a swine model, Cry1Ab-specific antibodies were not detected in pregnant sows or their offspring and no negative effects from feeding Bt maize to pregnant sows were observed.

In January 2013, the European Food Safety Authority released all data submitted by Monsanto in relation to the 2003 authorisation of maize genetically modified for glyphosate tolerance.

Starlink Corn Recalls

StarLink contains Cry9C, which had not previously been used in a GM crop. Starlink's creator, Plant Genetic Systems had applied to the US Environmental Protection Agency (EPA) to market Starlink for use in animal feed and in human food. However, because the Cry9C protein lasts longer in the digestive system than other Bt proteins, the EPA had concerns about its allergenicity, and PGS did not provide sufficient data to prove that Cry9C was not allergenic.[:3] As a result, PGS split its application into separate permits for use in food and use in animal feed. Starlink was approved by the EPA for use in animal feed only in May 1998.

StarLink corn was subsequently found in food destined for consumption by humans in the US, Japan, and South Korea. This corn became the subject of the widely publicized Starlink corn recall, which started when Taco Bell-branded taco shells sold in supermarkets were found to contain the corn. Sales of StarLink seed were discontinued. The registration for Starlink varieties was voluntarily withdrawn by Aventis in October 2000. (Pioneer had been bought by AgrEvo which then became Aventis CropScience at the time of the incident, which was later bought by Bayer

Fifty-one people reported adverse effects to the FDA; US Centers for Disease Control (CDC), which determined that 28 of them were possibly related to Starlink. However, the CDC studied the blood of these 28 individuals and concluded there was no evidence of hypersensitivity to the Starlink Bt protein.

A subsequent review of these tests by the Federal Insecticide, Fungicide, and Rodenticide Act Scientific Advisory Panel points out that while "the negative results decrease the probability that the Cry9C protein is the cause of allergic symptoms in the individuals examined ... in the absence of a positive control and questions regarding the sensitivity and specificity of the assay, it is not possible to assign a negative predictive value to this."

The US corn supply has been monitored for the presence of the Starlink Bt proteins since 2001.

In 2005, aid sent by the UN and the US to Central American nations also contained some StarLink corn. The nations involved, Nicaragua, Honduras, El Salvador and Guatemala refused to accept the aid.

Corporate Espionage

On December 19, 2013 six Chinese citizens were indicted in Iowa on charges of plotting to steal genetically modified seeds worth tens of millions of dollars from Monsanto and DuPont. Mo Hailong, director of international business at the Beijing Dabeinong Technology Group Co., part of the Beijing-based DBN Group, was accused of stealing trade secrets after he was found digging in an Iowa cornfield.

References

- Martins VAP (2008). "Genomic Insights into Oil Biodegradation in Marine Systems". Microbial Biodegradation: Genomics and Molecular Biology. Caister Academic Press. ISBN 978-1-904455-17-2.
- Langston, Jennifer (2016-11-22). "New grasses neutralize toxic pollution from bombs, explosives, and munitions". ScienceDaily. Retrieved 2016-11-30.
- "Infographics: Global Status of Commercialized Biotech/GM Crops: 2014 - ISAAA Brief 49-2014 | ISAAA.org". www.isaaa.org. Retrieved 2016-02-11.
- Pollack, Andrew (2007-11-27). "Round 2 for Biotech Beets". The New York Times. ISSN 0362-4331. Retrieved 2016-02-15.
- "Executive Summary: Global Status of Commercialized Biotech/GM Crops: 2014 - ISAAA Brief 49-2014 | ISAAA.org". www.isaaa.org. Retrieved 2016-02-16.
- Press, Associated (2010-03-03). "GM potato to be grown in Europe". The Guardian. ISSN 0261-3077. Retrieved 2016-02-15.
- "Pocket K No. 16: Global Status of Commercialized Biotech/GM Crops in 2014". isaaa.org. International Ser-

vice for the Acquisition of Agri-biotech Applications. Retrieved 23 February 2016.

- Final Report of the PABE research project (December 2001). "Public Perceptions of Agricultural Biotechnologies in Europe". Commission of European Communities. Retrieved February 24, 2016.
- Bashshur, Ramona (February 2013). "FDA and Regulation of GMOs". American Bar Association. Retrieved February 24, 2016.
- Lynch, Diahanna; Vogel, David (April 5, 2001). "The Regulation of GMOs in Europe and the United States: A Case-Study of Contemporary European Regulatory Politics". Council on Foreign Relations. Retrieved February 24, 2016.
- Final Report of the PABE research project (December 2001). "Public Perceptions of Agricultural Biotechnologies in Europe". Commission of European Communities. Retrieved February 24, 2016.
- Bashshur, Ramona (February 2013). "FDA and Regulation of GMOs". American Bar Association. Retrieved February 24, 2016.
- Lynch, Diahanna; Vogel, David (April 5, 2001). "The Regulation of GMOs in Europe and the United States: A Case-Study of Contemporary European Regulatory Politics". Council on Foreign Relations. Retrieved February 24, 2016.

6

Seed Producing Plants

Embryophyte is the most common group of plants that consist the vegetation on Earth. Embryophytes include liverworts, mosses, flowering plants and lycophytes. They are also referred to land plants majorly because of they live in terrestrial habitats. The section serves as a source to understand the main seeds that produce plants.

Embryophyte

The Embryophyta are the most familiar group of green plants that form vegetation on earth. Living embryophytes include hornworts, liverworts, mosses, ferns, lycophytes, gymnosperms and flowering plants, and emerged from Charophyte green algae. The Embryophyta are informally called land plants because they live primarily in terrestrial habitats, while the related green algae are primarily aquatic. All are complex multicellular eukaryotes with specialized reproductive organs. The name derives from their innovative characteristic of nurturing the young embryo sporophyte during the early stages of its multicellular development within the tissues of the parent gametophyte. With very few exceptions, embryophytes obtain their energy by photosynthesis, that is by using the energy of sunlight to synthesize their food from carbon dioxide and water.

Description

The evolutionary origins of the embryophytes are discussed further below, but they are believed to have evolved from within a group of complex green algae during the Paleozoic era (which started around 540 million years ago). Charales or the stoneworts may be the best living illustration of that developmental step. Embryophytes are primarily adapted for life on land, although some are secondarily aquatic. Accordingly, they are often called *land plants* or *terrestrial plants*.

On a microscopic level, the cells of embryophytes are broadly similar to those of green algae, but differ in that in cell division the daughter nuclei are separated by a phragmoplast. They are eukaryotic, with a cell wall composed of cellulose and plastids surrounded by two membranes. The latter include chloroplasts, which conduct photosynthesis and store food in the form of starch, and are characteristically pigmented with chlorophylls *a* and *b*, generally giving them a bright green color. Embryophyte cells also generally have an enlarged central vacuole enclosed by a vacuolar membrane or tonoplast, which maintains cell turgor and keeps the plant rigid.

In common with all groups of multicellular algae they have a life cycle which involves 'alternation of generations'. A multicellular generation with a single set of chromosomes – the haploid gametophyte – produces sperm and eggs which fuse and grow into a multicellular generation with twice the number of chromosomes – the diploid sporophyte. The mature sporophyte produces haploid spores which grow into a gametophyte, thus completing the cycle. Embryophytes have two features related to their reproductive cycles which distinguish them from all other plant lineages. Firstly, their gametophytes produce sperm and eggs in multicellular structures (called 'antheridia' and 'archegonia'), and fertilization of the ovum takes place within the archegonium rather than in the external environment. Secondly, and most importantly, the initial stage of development of the fertilized egg (the zygote) into a diploid multicellular sporophyte, take place within the archegonium where it is both protected and provided with nutrition. This second feature is the origin of the term 'embryophyte' – the fertilized egg develops into a protected embryo, rather than dispersing as a single cell. In the bryophytes the sporophyte remains dependent on the gametophyte, while in all other embryophytes the sporophyte generation is dominant and capable of independent existence.

Embryophytes also differ from algae by having metamers. Metamers are repeated units of development, in which each unit derives from a single cell, but the resulting product tissue or part is largely the same for each cell. The whole organism is thus constructed from similar, repeating parts or *metamers*. Accordingly, these plants are sometimes termed 'metaphytes' and classified as the group Metaphyta (but Haeckel's definition of Metaphyta places some algae in this group). In all land plants a disc-like structure called a phragmoplast forms where the cell will divide, a trait only found in the land plants in the streptophyte lineage, some species within their relatives Coleochaetales, Charales and Zygnematales, as well as within subaerial species of the algae order Trentepohliales, and appears to be essential in the adaptation towards a terrestrial life style.

Phylogeny and Classification

All green algae and land plants are now known to form a single evolutionary lineage or clade, one name for which is Viridiplantae (i.e. 'green plants'). According to several molecular clock estimates the Viridiplantae split 1,200 million years ago to 725 million years ago into two clades: chlorophytes and streptophytes. The chlorophytes are considerably more diverse (with around 700 genera) and were originally marine, although some groups have since spread into fresh water. The streptophyte algae (i.e. the streptophyte clade minus the land plants) are less diverse (with around 122 genera) and adapted to fresh water very early in their evolutionary history. They have not spread into marine environments (only a few stoneworts, which belong to this group, tolerate brackish water). Some time during the Ordovician period (which started around 490 million years ago) one or more streptophytes invaded the land and began the evolution of the embryophyte land plants.

Becker and Marin speculate that land plants evolved from streptophytes rather than any other group of algae because streptophytes were adapted to living in fresh water. This prepared them to tolerate a range of environmental conditions found on land. Fresh water living made them tolerant of exposure to rain; living in shallow pools required tolerance to temperature variation, high levels of ultra-violet light and seasonal dehydration.

Relationships between the groups making up Viridiplantae are still being elucidated. Views have changed considerably since 2000 and classifications have not yet caught up. However, the division between chlorophytes and streptophytes and the evolution of embryophytes from within the latter group, as shown in the cladogram below, are well established. Three approaches to classification are shown. Older classifications, as on the left, treated all green algae as a single division of the plant kingdom under the name Chlorophyta. Land plants were then placed in separate divisions. All the streptophyte algae can be grouped into one paraphyletic taxon, as in the middle, allowing the embryophytes to form a taxon at the same level. Alternatively, the embryophytes can be sunk into a monophyletic taxon comprising all the streptophytes, as shown below. A variety of names have been used for the different groups which result from these approaches; those used below are only one of a number of possibilities. The higher-level classification of the Viridiplantae varies considerably, resulting in widely different ranks being assigned to the embryophytes, from kingdom to class.

The precise relationships within the streptophytes are less clear as of March 2012. The stoneworts (Charales) have traditionally been identified as closest to the embryophytes, but recent work suggests that either the Zygnematales or a clade consisting of the Zygnematales and the Coleochaetales may be the sister group to the land plants. That the Zygnematales (or Zygnematophyceae) are the closest algal relatives to land plants was underpinned by an exhaustive phylogenetic analysis (phylogenomics) performed in 2014, which is supported by both plastid genome phylogenies as well as plastid gene content and properties.

The preponderance of currently available molecular evidence suggests that the groups making up the embryophytes are related as shown in the cladogram below (based on Qiu *et al.* 2006 with additional names from Crane *et al.* 2004).

Studies based on morphology rather than on genes and proteins have regularly reached different conclusions; for example that neither the monilophytes (ferns and horsetails) nor the gymnosperms are a natural or monophyletic group.

There is considerable variation in how these relationships are converted into a formal classification. Consider the angiosperms or flowering plants. Many botanists, following Lindley in 1830, have treated the angiosperms as a division. Researchers concerned with fossil plants have usually followed Banks in treating the tracheophytes or vascular plants as a division, so that the angiosperms become a class or even a subclass. Two very different systems are shown below. The classification on the left is a traditional one, in which ten living groups are treated as separate divisions; the classification on the right (based on Kenrick and Crane's 1997 treatment) sharply reduces the rank of groups such as the flowering plants. (More complex classifications are needed if extinct plants are included.)

Two contrasting classifications of living land plants		
Liverworts	Marchiantiophyta	Marchiantiophyta
Mosses	Bryophyta	Bryophyta
Hornworts	Anthocerotophyta	Anthocerotophyta
		Tracheophyta
Lycophytes	Lycopodiophyta	Lycophytina
		Euphyllophytina
Ferns and horsetails	Pteridophyta	Moniliformopses
		Radiatopses
Cycads	Cycadophyta	Cycadatae
Conifers	Pinophyta	Coniferophytatae

Ginkgo	Ginkgophyta	Ginkgoatae
Gnetophytes	Gnetophyta	Anthophytatae
Flowering plants	Magnoliophyta	

Diversity

Bryophytes

Most bryophytes, such as these mosses, produce stalked sporophytes from which their spores are released.

Bryophytes consist of all non-vascular land plants (embryophytes without vascular tissue). All are relatively small and are usually confined to environments that are humid or at least seasonally moist. They are limited by their reliance on water needed to disperse their gametes, although only a few bryophytes are truly aquatic. Most species are tropical, but there are many arctic species as well. They may locally dominate the ground cover in tundra and Arctic–alpine habitats or the epiphyte flora in rain forest habitats.

The three living divisions are the mosses (Bryophyta), hornworts (Anthocerotophyta), and liverworts (Marchantiophyta). Originally, these three groups were included together as classes within the single division Bryophyta. However, they now are placed separately into three divisions since the bryophytes as a whole are known to be a paraphyletic (artificial) group instead of a single lineage. Instead, the three bryophyte groups form an evolutionary grade of those land plants that are not vascular. Some closely related green algae are also non-vascular, but are not considered "land plants."

- Marchantiophyta (Liverworts)
- Bryophyta (mosses)
- Anthocerotophyta (hornworts)

Despite the fact that they are no longer classified as a single group, the bryophytes are still studied together because of their many biological similarities as non-vascular land plants. All three bryophyte groups share a haploid-dominant life cycle and unbranched sporophytes. These are traits that appear to be plesiotypic within the land plants, and thus were common to all early diverging lineages of plants on the land. The fact that the bryophytes have a life cycle in common is thus an artefact of being the oldest extant lineages of land plant, and not the result of close shared ancestry.

The bryophyte life-cycle is strongly dominated by the haploid gametophyte generation. The sporophyte remains small and dependent on the parent gametophyte for its entire brief life. All other living groups of land plants have a life cycle dominated by the diploid sporophyte generation. It is in the diploid sporophyte that vascular tissue develops. Although some mosses have quite complex water-conducting vessels, bryophytes lack true vascular tissue.

Like the vascular plants, bryophytes do have differentiated stems, and although these are most often no more than a few centimeters tall, they do provide mechanical support. Most bryophytes also have leaves, although these typically are one cell thick and lack veins. Unlike the vascular plants, bryophytes lack true roots or any deep anchoring structures. Some species do grow a filamentous network of horizontal stems, but these have a primary function of mechanical attachment rather than extraction of soil nutrients (Palaeos 2008).

Rise of Vascular Plants

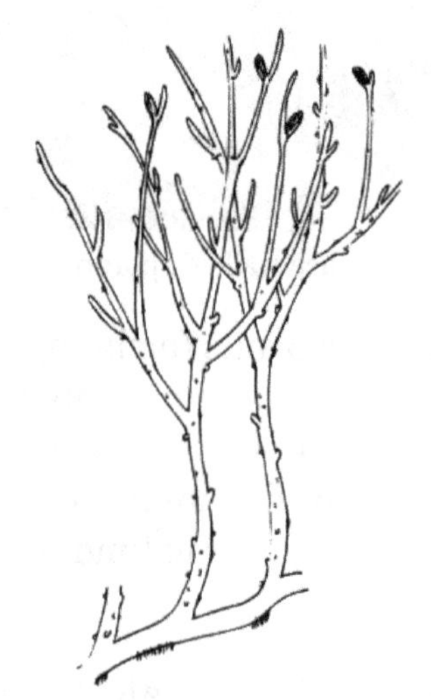

Reconstruction of a plant of *Rhynia*

During the Silurian and Devonian periods (around 440 to 360 million years ago), plants evolved which possessed true vascular tissue, including cells with walls strengthened by lignin (tracheids). Some extinct early plants appear to be between the grade of organization bryophytes and that of true vascular plants (eutracheophytes). Genera such as *Horneophyton* have water-conducting tissue more like that of mosses, but a different life-cycle in which the sporophyte is more developed than the gametophyte. Genera such as *Rhynia* have a similar life-cycle but have simple tracheids and so are a kind of vascular plant.

During the Devonian period, vascular plants diversified and spread to many different land environments. In addition to vascular tissues which transport water throughout the body, tracheophytes have an outer layer or cuticle that resists drying out. The sporophyte is the dominant generation, and in modern species develops leaves, stems and roots, while the gametophyte remains very small.

Lycophytes and Euphyllophytes

All the vascular plants which disperse through spores were once thought to be related (and were often grouped as 'ferns and allies'). However, recent research suggests that leaves evolved quite separately in two different lineages. The lycophytes or lycopodiophytes – modern clubmosses, spikemosses and quillworts – make up less than 1% of living vascular plants. They have small leaves, often called 'microphylls' or 'lycophylls', which are borne all along the stems in the club-mosses and spikemosses, and which effectively grow from the base, via an intercalary meristem. It is believed that microphylls evolved from outgrowths on stems, such as spines, which later acquired veins (vascular traces).

Lycopodiella inundata, a lycophyte

Although the living lycophytes are all relatively small and inconspicuous plants, more common in the moist tropics than in temperate regions, during the Carboniferous period tree-like lycophytes (such as *Lepidodendron*) formed huge forests that dominated the landscape.

The euphyllophytes, making up more than 99% of living vascular plant species, have large 'true' leaves (megaphylls), which effectively grow from the sides or the apex, via marginal or apical meristems. One theory is that megaphylls developed from three-dimensional branching systems by first 'planation' – flattening to produce a two dimensional branched structure – and then 'webbing' – tissue growing out between the flattened branches. Others have questioned whether megaphylls developed in the same way in different groups.

Ferns and Horsetails

Athyrium filix-femina, unrolling young frond

Euphyllophytes are divided into two lineages: the ferns and horsetails (monilophytes) and the seed plants (spermatophytes). Like all the preceding groups, the monilophytes continue to use spores as their main method of dispersal. Traditionally, whisk ferns and horsetails were treated as distinct from 'true' ferns. Recent research suggests that they all belong together, although there are differences of opinion on the exact classification to be used. Living whisk ferns and horsetails do not have the large leaves (megaphylls) which would be expected of euphyllophytes. However, this has probably resulted from reduction, as evidenced by early fossil horsetails, in which the leaves are broad with branching veins.

Ferns are a large and diverse group, with some 12,000 species. A stereotypical fern has broad, much divided leaves, which grow by unrolling.

Seed Plants

Conifer forest in Northern California

Large seed of a horse chestnut, *Aesculus hippocastanum*

Seed plants, which first appeared in the fossil record towards the end of the Paleozoic era, reproduce using desiccation-resistant capsules called seeds. Starting from a plant which disperses by spores, highly complex changes are needed to produce seeds. The sporophyte has two kinds of spore-forming organs (sporangia). One kind, the megasporangium, produces only a single large spore (a megaspore). This sporangium is surrounded by one or more sheathing layers (integuments) which form the seed coat. Within the seed coat, the megaspore develops into a tiny gametophyte, which in turn produces one or more egg cells. Before fertilization, the sporangium and its contents plus its coat is called an 'ovule'; after fertilization a 'seed'. In parallel to these developments, the other kind of sporangium, the microsporangium, produces microspores. A tiny gametophyte develops inside the wall of a microspore, producing a pollen grain. Pollen grains are physically transferred between plants by the wind or animals, most commonly insects. When a pollen grain reaches an ovule, it enters via a microscopic gap in the coat (the micropyle). The tiny gametophyte inside the pollen grain then produces sperm cells which move to the egg cell and fertilize it. Seed plants include two groups with living members, the gymnosperms and the angiosperms or flowering plants. In gymnosperms, the ovules or seeds are not further enclosed. In angiosperms, they are enclosed in ovaries. A split ovary with a visible seed can be seen in the adjacent image. Angiosperms typically also have other, secondary structures, such as petals, which together form a flower.

Extant seed plants are divided into five groups:

Gymnosperms

- Pinophyta - conifers
- Cycadophyta - cycads
- Ginkgophyta - ginkgo
- Gnetophyta - gnetophytes

Angiosperms

- Magnoliophyta – flowering plants

Spermatophyte

The spermatophytes, also known as phanerogams or phenogamae, comprise those plants that produce seeds, hence the alternative name seed plants. They are a subset of the embryophytes or land plants. These terms distinguished those plants with hidden sexual organs (cryptogamae) from those with visible sexual organs (phanerogamae).

Description

The living spermatophytes form five groups:

- cycads, a subtropical and tropical group of plants with a large crown of compound leaves and a stout trunk,
- *ginkgo*, a single living species of tree,
- conifers, cone-bearing trees and shrubs,
- gnetophytes, woody plants in the genera *Ephedra*, *Gnetum*, and *Welwitschia*
- angiosperms, (or magnoliophyta) the flowering plants, a large group including many familiar plants in a wide variety of habitats.

In addition to the taxa listed above, the fossil (old creature) record contains evidence of many extinct taxa of seed plants. The so-called "seed ferns" (Pteridospermae) were one of the earliest successful groups of land plants, and forests dominated by seed ferns were prevalent in the late Paleozoic. *Glossopteris* was the most prominent tree genus in the ancient southern supercontinent of Gondwana during the Permian period. By the Triassic period, seed ferns had declined in ecological importance, and representatives of modern gymnosperm groups were abundant and dominant through the end of the Cretaceous, when angiosperms radiated.

Evolution

A whole genome duplication event in the ancestor of seed plants occurred about 319 million years ago. This gave rise to a series of evolutionary changes that resulted in the origin of seed plants.

A middle Devonian precursor to seed plants from Belgium has been identified predating the earliest seed plants by about 20 million years. *Runcaria*, small and radially symmetrical, is an integumented megasporangium surrounded by a cupule. The megasporangium bears an unopened distal extension protruding above the mutlilobed integument. It is suspected that the extension was involved in anemophilous pollination. *Runcaria* sheds new light on the sequence of character acquisition leading to the seed. *Runcaria* has all of the qualities of seed plants except for a solid seed coat and a system to guide the pollen to the seed.

Relationships and Nomenclature

Seed-bearing plants were traditionally divided into angiosperms, or flowering plants, and gymnosperms, which includes the gnetophytes, cycads, ginkgo, and conifers. Older morphological studies believed in a close relationship between the gnetophytes and the angiosperms, in particular based on vessel elements. However, molecular studies (and some more recent morphological and fossil papers) have generally shown a clade of gymnosperms, with the gnetophytes in or near the conifers. For example, one common proposed set of relationships is known as the *gne-pine hypothesis* and looks like:

However, the relationships between these groups should not be considered settled.

Other classifications group all the seed plants in a single division, with classes for our five groups:

- Division Spermatophyta
    - Cycadopsida, the cycads
    - Ginkgoopsida, the ginkgo
    - Pinopsida, the conifers, ("Coniferopsida")
    - Gnetopsida, the gnetophytes
    - Magnoliopsida, the flowering plants, or Angiospermopsida

A more modern classification ranks these groups as separate divisions (sometimes under the Superdivision Spermatophyta):

- Cycadophyta, the cycads
- Ginkgophyta, the ginkgo
- Pinophyta, the conifers

- Gnetophyta, the gnetophytes
- Magnoliophyta, the flowering plants

Ovule

Location of ovules inside a *Helleborus foetidus* flower

In seed plants, the ovule ("small egg") is the structure that gives rise to and contains the female reproductive cells. It consists of three parts: The integument(s) forming its outer layer(s), the nucellus (or remnant of the megasporangium), and female gametophyte (formed from haploid megaspore) in its center. The female gametophyte—specifically termed a *megagametophyte*—is also called the embryo sac in angiosperms. The megagametophyte produces an egg cell (or several egg cells in some groups) for the purpose of fertilization. After fertilization, the ovule develops into a seed.

Location within the Plant

In flowering plants, the ovule is located inside the portion of the flower called the gynoecium. The ovary of the gynoecium produces one or more ovules and ultimately becomes the fruit wall. Ovules are attached to the placenta in the ovary through a stalk-like structure known as a funiculus (plural, funiculi). Different patterns of ovule attachment, or placentation, can be found among plant species, these include:

- Apical placentation: The placenta is at the apex (top) of the ovary. Simple or compound ovary.
- Axile placentation: The ovary is divided into radial segments, with placentas in separate locules. Ventral sutures of carpels meet at the centre of the ovary. Placentae are along fused margins of carpels. Two or more carpels. (e.g. *Hibiscus*, *Citrus*, *Solanum*)
- Basal placentation: The placenta is at the base (bottom) of the ovary on a protrusion of the thalamus (receptacle). Simple or compound carpel, unilocular ovary. (e.g. *Sonchus*, *Helianthus*, Compositae)

- Free-central placentation: Derived from axile as partitions are absorbed, leaving ovules at the central axis. Compound unilocular ovary. (e.g. *Stellaria*, *Dianthus*)
- Marginal placentation: Simplest type. There is only one elongated placenta on one side of the ovary, as ovules are attached at the fusion line of the carpel's margins . This is conspicuous in legumes. Simple carpel, unilocular ovary. (e.g. *Pisum*)
- Parietal placentation: Placentae on inner ovary wall within a non-sectioned ovary, corresponding to fused carpel margins. Two or more carpels, unilocular ovary. (e.g. *Brassica*)
- Superficial: Similar to axile, but placentae are on inner surfaces of multilocular ovary (e.g. *Nymphaea*)

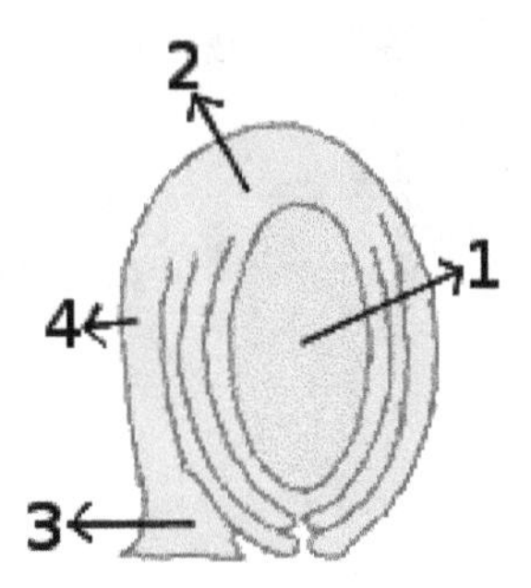

Ovule structure (anatropous) 1: nucleus 2: chalaza 3: funiculus 4: raphe

Ovule orientation may be anatropous, such that when inverted the micropyle faces the placenta (this is the most common ovule orientation in flowering plants), amphitropous, campylotropous, or orthotropous.

In gymnosperms such as conifers, ovules are borne on the surface of an ovuliferous (ovule-bearing) scale, usually within an ovulate cone (also called megastrobilus). In some extinct plants (e.g. Pteridosperms), megasporangia and perhaps ovules were borne on the surface of leaves. In other extinct taxa, a cupule (a modified leaf or part of a leaf) surrounds the ovule (e.g. *Caytonia* or *Glossopteris*).

Ovule Parts and Development

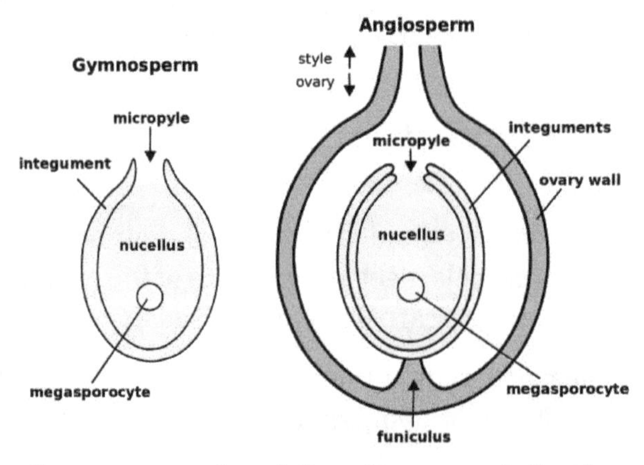

Plant ovules: Gymnosperm ovule on left, angiosperm ovule (inside ovary) on right

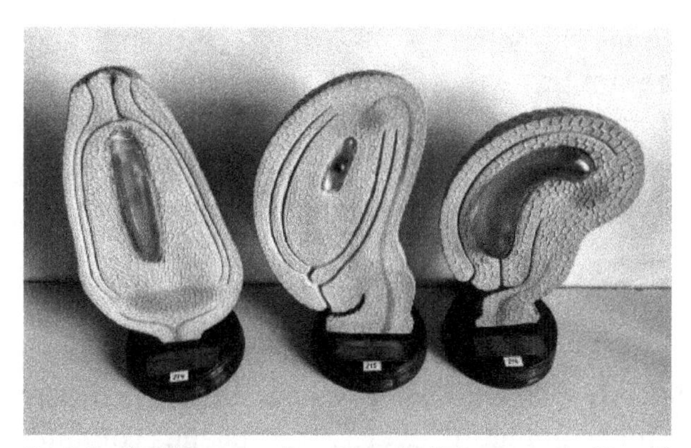
Models of different ovules, Botanical Museum Greifswald

The ovule appears to be a megasporangium with integuments surrounding it. Ovules are initially composed of diploid maternal tissue, which includes a megasporocyte (a cell that will undergo meiosis to produce megaspores). Megaspores remain inside the ovule and divide by mitosis to produce the haploid female gametophyte or megagametophyte, which also remains inside the ovule. The remnants of the megasporangium tissue (the nucellus) surround the megagametophyte. Megagametophytes produce archegonia (lost in some groups such as flowering plants), which produce egg cells. After fertilization, the ovule contains a diploid zygote and then, after cell division begins, an embryo of the next sporophyte generation. In flowering plants, a second sperm nucleus fuses with other nuclei in the megagametophyte forming a typically polyploid (often triploid) endosperm tissue, which serves as nourishment for the young sporophyte.

Integuments, Micropyle and Chalaza

An integument is a protective cell layer surrounding the ovule. Gymnosperms typically have one integument (unitegmic) while angiosperms typically have two (bitegmic). The evolutionary origin of the inner integument (which is integral to the formation of ovules from megasporangia) has been proposed to be by enclosure of a megasporangium by sterile branches (telomes). *Elkinsia*, a preovulate taxon, has a lobed structure fused to the lower third of the megasporangium, with the lobes extending upwards in a ring around the megasporangium. This might, through fusion between lobes and between the structure and the megasporangium, have produced an integument.

The origin of the second or outer integument has been an area of active contention for some time. The cupules of some extinct taxa have been suggested as the origin of the outer integument. A few angiosperms produce vascular tissue in the outer integument, the orientation of which suggests that the outer surface is morphologically abaxial. This suggests that cupules of the kind produced by the Caytoniales or Glossopteridales may have evolved into the outer integument of angiosperms.

The integuments develop into the seed coat when the ovule matures after fertilization.

The integuments do not enclose the nucellus completely but retain an opening at the apex referred to as the micropyle. The micropyle opening allows the pollen (a male gametophyte) to enter the ovule for fertilization. In gymnosperms (e.g., conifers), the pollen is drawn into the ovule on a drop of fluid that exudes out of the micropyle, the so-called pollination drop mechanism. Subsequently, the micropyle closes. In angiosperms, only a pollen tube enters the micropyle. During germination, the seedling's radicle emerges through the micropyle.

Located opposite from the micropyle is the chalaza where the nucellus is joined to the integuments. Nutrients from the plant travel through the phloem of the vascular system to the funiculus and outer integument and from there apoplastically and symplastically through the chalaza to the nucellus inside the ovule. In chalazogamous plants, the pollen tubes enter the ovule through the chalaza instead of the micropyle opening.

Nucellus, Megaspore and Perisperm

The nucellus (plural: nucelli) is part of the inner structure of the ovule, forming a layer of diploid (sporophytic) cells immediately inside the integuments. It is structurally and functionally equivalent to the megasporangium. In immature ovules, the nucellus contains a megasporocyte (megaspore mother cell), which undergoes sporogenesis via meiosis. In gymnosperms, three of the four haploid spores produced in meiosis typically degenerate, leaving one surviving megaspore inside the nucellus. Among angiosperms, however, a wide range of variation exists in what happens next. The number (and position) of surviving megaspores, the total number of cell divisions, whether nuclear fusions occur, and the final number, position and ploidy of the cells or nuclei all vary. A common pattern of embryo sac development (the *Polygonum* type maturation pattern) includes a single functional megaspore followed by three rounds of mitosis. In some cases, however, two megaspores survive (for example, in *Allium* and *Endymion*). In some cases all four megaspores survive, for example in the *Fritillaria* type of development (illustrated by *Lilium* in the figure) there is no separation of the megaspores following meiosis, then the nuclei fuse to form a triploid nucleus and a haploid nucleus. The subsequent arrangement of cells is similar to the *Polygonum* pattern, but the ploidy of the nuclei is different.

After fertilization, the nucellus may develop into the perisperm that feeds the embryo. In some plants, the diploid tissue of the nucellus can give rise to the embryo within the seed through a mechanism of asexual reproduction called nucellar embryony.

Megagametophyte

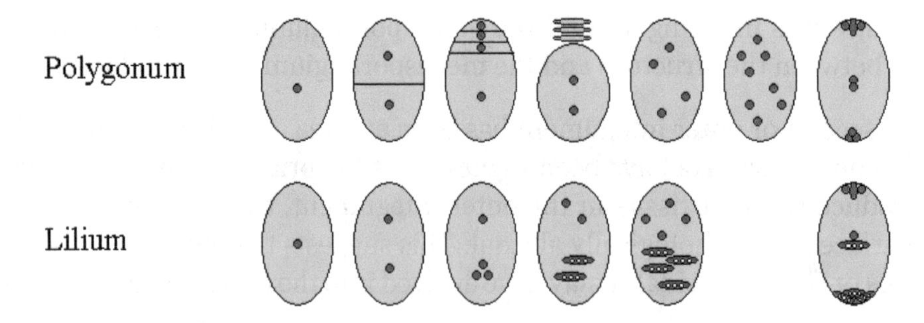

Megagametophyte formation of the genera *Polygonum* and *Lilium*. Triploid nuclei are shown as ellipses with three white dots. The first three columns show the meiosis of the megaspore, followed by 1-2 mitoses.

The haploid megaspore inside the nucellus gives rise to the female gametophyte, called the megagametophyte.

In gymnosperms, the megagametophyte consists of around 2000 nuclei and forms archegonia, which produce egg cells for fertilization.

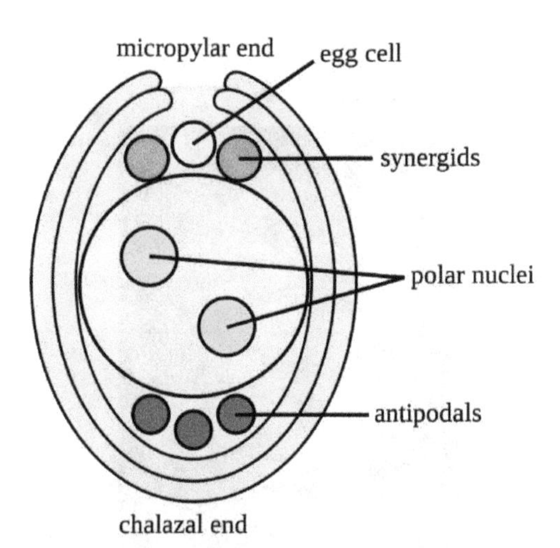

Ovule with megagametophyte: egg cell (yellow), synergids (orange), central cell with two polar nuclei (bright green), and antipodals (dark green)

In flowering plants, the megagametophyte (also referred to as the embryo sac) is much smaller and typically consists of only seven cells and eight nuclei. This type of megagametophyte develops from the megaspore through three rounds of mitotic divisions. The cell closest to the micropyle opening of the integuments differentiates into the egg cell, with two synergid cells by its side that are involved in the production of signals that guide the pollen tube. Three antipodal cells form on the opposite (chalazal) end of the ovule and later degenerate. The large central cell of the embryo sac contains two polar nuclei.

Zygote, Embryo and Endosperm

The pollen tube releases two sperm nuclei into the ovule. In gymnosperms, fertilization occurs within the archegonia produced by the female gametophyte. While it is possible that several egg cells are present and fertilized, typically only one zygote will develop into a mature embryo as the resources within the seed are limited.

In flowering plants, one sperm nucleus fuses with the egg cell to produce a zygote, the other fuses with the two polar nuclei of the central cell to give rise to the polyploid (typically triploid) endosperm. This double fertilization is unique to flowering plants, although in some other groups the second sperm cell does fuse with another cell in the megagametophyte to produce a second embryo. The plant stores nutrients such as starch, proteins, and oils in the endosperm as a food source for the developing embryo and seedling, serving a similar function to the yolk of animal eggs. The endosperm is also called the albumen of the seed.

Embryos may be described by a number of terms including Linear (embryos have axile placentation and are longer than broad), or rudimentary (embryos are basal in which the embryo is tiny in relation to the endosperm).

Types of Gametophytes

Megagametophytes of flowering plants may be described according to the number of megaspores developing, as either monosporic, bisporic, or tetrasporic.

Gymnosperm

Encephalartos sclavoi cone, about 30 cm long

The gymnosperms are a group of seed-producing plants that includes conifers, cycads, *Ginkgo*, and gnetophytes. Their naked condition stands in contrast to the seeds and ovules of flowering plants (angiosperms), which are enclosed within an ovary. Gymnosperm seeds develop either on the surface of scales or leaves, often modified to form cones, or at the end of short stalks as in *Ginkgo*.

The gymnosperms and angiosperms together compose the spermatophytes or seed plants. By far the largest group of living gymnosperms are the conifers (pines, cypresses, and relatives), followed by cycads, gnetophytes (*Gnetum*, *Ephedra* and *Welwitschia*), and *Ginkgo* (a single living species).

Classification

In early classification schemes, the gymnosperms (Gymnospermae) were regarded as a "natural" group. There is conflicting evidence on the question of whether the living gymnosperms form a clade. The fossil record of gymnosperms includes many distinctive taxa that do not belong to the four modern groups, including seed-bearing trees that have a somewhat fern-like vegetative morphology (the so-called "seed ferns" or pteridosperms.) When fossil gymnosperms such as Bennettitales, *Caytonia* and the glossopterids are considered, it is clear that angiosperms are nested within a larger gymnosperm clade, although which group of gymnosperms is their closest relative remains unclear.

For the most recent classification on extant gymnosperms see Christenhusz *et al.* (2011). There are 12 families, 83 known genera with a total of ca 1080 known species (Christenhusz & Byng 2016).

Subclass Cycadidae

- Order Cycadales
    - Family Cycadaceae: *Cycas*
    - Family Zamiaceae: *Dioon, Bowenia, Macrozamia, Lepidozamia, Encephalartos, Stangeria, Ceratozamia, Microcycas, Zamia.*

Subclass Ginkgoidae

- Order Ginkgoales
    - Family Ginkgoaceae: *Ginkgo*

Subclass Gnetidae

- Order Welwitschiales
    - Family Welwitschiaceae: *Welwitschia*
- Order Gnetales
    - Family Gnetaceae: *Gnetum*
- Order Ephedrales
    - Family Ephedraceae: *Ephedra*

Subclass Pinidae

- Order Pinales
    - Family Pinaceae: *Cedrus, Pinus, Cathaya, Picea, Pseudotsuga, Larix, Pseudolarix, Tsuga, Nothotsuga, Keteleeria, Abies*
- Order Araucariales
    - Family Araucariaceae: *Araucaria, Wollemia, Agathis*
    - Family Podocarpaceae: *Phyllocladus, Lepidothamnus, Prumnopitys, Sundacarpus, Halocarpus, Parasitaxus, Lagarostrobos, Manoao, Saxegothaea, Microcachrys, Pherosphaera, Acmopyle, Dacrycarpus, Dacrydium, Falcatifolium, Retrophyllum, Nageia, Afrocarpus, Podocarpus*
- Order Cupressales
    - Family Sciadopityaceae: *Sciadopitys*
    - Family Cupressaceae: *Cunninghamia, Taiwania, Athrotaxis, Metasequoia, Sequoia, Sequoiadendron, Cryptomeria, Glyptostrobus, Taxodium, Papuacedrus, Austrocedrus, Libocedrus, Pilgerodendron, Widdringtonia, Diselma, Fitzroya, Callitris* (incl. *Actinostrobus* and *Neocallitropsis*), *Thujopsis, Thuja, Fokienia,*

Chamaecyparis, Callitropsis, Cupressus, Juniperus, Xanthocyparis, Calocedrus, Tetraclinis, Platycladus, Microbiota

- Family Taxaceae: *Austrotaxus, Pseudotaxus, Taxus, Cephalotaxus, Amentotaxus, Torreya*

Diversity and Origin

There are more than 1000 extant or currently living species of gymnosperms in 88 plant genera belonging to 14 plant families.

It is widely accepted that the gymnosperms originated in the late Carboniferous period, replacing the lycopsid rainforests of the tropical region. This appears to have been the result of a whole genome duplication event around 319 million years ago. Early characteristics of seed plants were evident in fossil progymnosperms of the late Devonian period around 380 million years ago. It has been suggested that during the mid-Mesozoic era, pollination of some extinct groups of gymnosperms was by extinct species of scorpionflies that had specialized proboscis for feeding on pollination drops. The scorpionflies likely engaged in pollination mutualisms with gymnosperms, long before the similar and independent coevolution of nectar-feeding insects on angiosperms. Evidence has also been found that mid-Mesozoic gymnosperms were pollinated by Kalligrammatid lacewings, a now-extinct genus with members which (in an example of convergent evolution) resembled the modern butterflies that arose far later.

Conifers are by far the most abundant extant group of gymnosperms with six to eight families, with a total of 65-70 genera and 600-630 species (696 accepted names). Conifers are woody plants and most are evergreens. The leaves of many conifers are long, thin and needle-like, other species, including most Cupressaceae and some Podocarpaceae, have flat, triangular scale-like leaves. *Agathis* in Araucariaceae and *Nageia* in Podocarpaceae have broad, flat strap-shaped leaves.

Cycads are the next most abundant group of gymnosperms, with two or three families, 11 genera, and approximately 300 species. The other extant groups are the 75-80 species of Gnetales and one species of Ginkgo.

Uses

Gymnosperms have major economic uses. Pine, fir, spruce, and cedar are all examples of conifers that are used for lumber. Some other common uses for gymnosperms are soap, varnish, nail polish, food, gum, and perfumes.

Life Cycle

Gymnosperms, like all vascular plants, have a sporophyte-dominant life cycle. The gametophyte (gamete-bearing phase) is relatively short-lived. Two spore types, microspores and megaspores, are typically produced in pollen cones or ovulate cones, respectively. Gametophytes, as with all heterosporous plants, develop within the spore wall. Pollen grains (microgametophytes) mature from microspores, and ultimately produce sperm cells. Megagametophytes develop from megaspores and are retained within the ovule. They typically produce multiple archegonia. During pollination, pollen grains are physically transferred between plants, from pollen cone to the ovule, being transferred by wind or insects. Whole grains enter each ovule through a microscopic gap in the ovule coat (integument) called the micropyle. The pollen grains mature further inside the ovule and produce sperm cells. Two main modes of fertilization are found in gymnosperms. Cycads and *Ginkgo* have motile sperm that swim directly to the egg inside the ovule, whereas conifers and gnetophytes have sperm with no flagella that are conveyed to the egg along a pollen tube. After syngamy (joining of the sperm and egg cell), the zygote develops into an embryo (young sporophyte). More than one embryo is usually initiated in each gymnosperm seed. The mature seed comprises the embryo and the remains of the female gametophyte, which serves as a food supply, and the seed coat (integument).

Genetics

The first published sequenced genome for any gymnosperm was the genome of *Picea abies* in 2013.

Flowering Plant

The flowering plants (angiosperms), also known as Angiospermae or Magnoliophyta, are the most diverse group of land plants, with 416 families, approx. 13,164 known genera and a total of c. 295,383 known species. Like gymnosperms, angiosperms are seed-producing plants; they are distinguished from gymnosperms by characteristics including flowers, endosperm within the seeds, and the production of fruits that contain the seeds. Etymologically, angiosperm means a plant that produces seeds within an enclosure, in other words, a fruiting plant. The term "angiosperm" comes from the Greek composite word (*angeion*, "case" or "casing", and *sperma*, "seed") meaning "enclosed seeds", after the enclosed condition of the seeds.

The ancestors of flowering plants diverged from gymnosperms in the Triassic Period, during the range 245 to 202 million years ago (mya), and the first flowering plants are known from 160 mya. They diversified extensively during the Lower Cretaceous, became widespread by 120 mya, and replaced conifers as the dominant trees from 100 to 60 mya.

Angiosperm Derived Characteristics

Bud of a pink rose

Angiosperms differ from other seed plants in several ways, described in the table. These distinguishing characteristics taken together have made the angiosperms the most diverse and numerous land plants and the most commercially important group to humans.

Distinctive features of Angiosperms	
Feature	Description
Flowering organs	Flowers, the reproductive organs of flowering plants, are the most remarkable feature distinguishing them from the other seed plants. Flowers provided angiosperms with the means to have a more species-specific breeding system, and hence a way to evolve more readily into different species without the risk of crossing back with related species. Faster speciation enabled the Angiosperms to adapt to a wider range of ecological niches. This has allowed flowering plants to largely dominate terrestrial ecosystems.
Stamens with two pairs of pollen sacs	Stamens are much lighter than the corresponding organs of gymnosperms and have contributed to the diversification of angiosperms through time with adaptations to specialized pollination syndromes, such as particular pollinators. Stamens have also become modified through time to prevent self-fertilization, which has permitted further diversification, allowing angiosperms eventually to fill more niches.
Reduced male parts, three cells	The male gametophyte in angiosperms is significantly reduced in size compared to those of gymnosperm seed plants. The smaller size of the pollen reduces the amount of time between pollination — the pollen grain reaching the female plant — and fertilization. In gymnosperms, fertilization can occur up to a year after pollination, whereas in angiosperms, fertilization begins very soon after pollination. The shorter amount of time between pollination and fertilization allows angiosperms to produce seeds earlier after pollination than gymnosperms, providing angiosperms a distinct evolutionary advantage.
Closed carpel enclosing the ovules (carpel or carpels and accessory parts may become the fruit)	The closed carpel of angiosperms also allows adaptations to specialized pollination syndromes and controls. This helps to prevent self-fertilization, thereby maintaining increased diversity. Once the ovary is fertilized, the carpel and some surrounding tissues develop into a fruit. This fruit often serves as an attractant to seed-dispersing animals. The resulting cooperative relationship presents another advantage to angiosperms in the process of dispersal.
Reduced female gametophyte, seven cells with eight nuclei	The reduced female gametophyte, like the reduced male gametophyte, may be an adaptation allowing for more rapid seed set, eventually leading to such flowering plant adaptations as annual herbaceous life-cycles, allowing the flowering plants to fill even more niches.
Endosperm	In general, endosperm formation begins after fertilization and before the first division of the zygote. Endosperm is a highly nutritive tissue that can provide food for the developing embryo, the cotyledons, and sometimes the seedling when it first appears.

Evolution

Fossilized spores suggest that higher plants (embryophytes) have lived on land for at least 475 million years. Early land plants reproduced sexually with flagellated, swimming sperm, like the green algae from which they evolved. An adaptation to terrestrialization was the development of upright meiosporangia for dispersal by spores to new habitats. This feature is lacking in the descendants of their nearest algal relatives, the Charophycean green algae. A later terrestrial adaptation took place with retention of the delicate, avascular sexual stage, the gametophyte, within the tissues of the vascular sporophyte. This occurred by spore germination within sporangia rather than spore release, as in non-seed plants. A current example of how this might have happened can be seen in the precocious spore germination in *Selaginella*, the spike-moss. The result for the ancestors of angiosperms was enclosing them in a case, the seed. The first seed bearing plants, like the ginkgo, and conifers (such as pines and firs), did not produce flowers. The pollen grains (males) of *Ginkgo* and cycads produce a pair of flagellated, mobile sperm cells that "swim" down the developing pollen tube to the female and her eggs.

Flowers of *Malus sylvestris* (crab apple)

Flowers and leaves of *Oxalis pes-caprae* (Bermuda buttercup)

The apparently sudden appearance of nearly modern flowers in the fossil record initially posed such a problem for the theory of evolution that Charles Darwin called it an "*abominable mystery*". However, the fossil record has considerably grown since the time of Darwin, and recently discovered angiosperm fossils such as *Archaefructus*, along with further discoveries of fossil gymnosperms, suggest how angiosperm characteristics may have been acquired in a series of steps. Several groups of extinct gymnosperms, in particular seed ferns, have been proposed as the ancestors of flowering plants, but there is no continuous fossil evidence showing exactly how flowers evolved. Some older fossils, such as the upper Triassic *Sanmiguelia*, have been suggested. Based on current evidence, some propose that the ancestors of the angiosperms diverged from an unknown group of gymnosperms in the Triassic period (245–202 million years ago). Fossil angiosperm-like pollen from the Middle Triassic (247.2–242.0 Ma) suggests an older date for their origin. A close relationship between angiosperms and gnetophytes, proposed on the basis of morphological evidence, has more recently been disputed on the basis of molecular evidence that suggest gnetophytes are instead more closely related to other gymnosperms.

The evolution of seed plants and later angiosperms appears to be the result of two distinct rounds of whole genome duplication events. These occurred at 319 million years ago and 192 million years ago. Another possible whole genome duplication event at 160 million years ago perhaps created the ancestral line that led to all modern flowering plants. That event was studied by sequencing the genome of an ancient flowering plant, *Amborella trichopoda,* and directly addresses Darwin's "*abominable mystery.*"

The earliest known macrofossil confidently identified as an angiosperm, *Archaefructus liaoningensis,* is dated to about 125 million years BP (the Cretaceous period), whereas pollen considered to be of angiosperm origin takes the fossil record back to about 130 million years BP. However, one study has suggested that the early-middle Jurassic plant *Schmeissneria,* traditionally considered a type of ginkgo, may be the earliest known angiosperm, or at least a close relative. In addition, circumstantial chemical evidence has been found for the existence of angiosperms as early as 250 million years ago. Oleanane, a secondary metabolite produced by many flowering plants, has been found in Permian deposits of that age together with fossils of gigantopterids. Gigantopterids are a group of extinct seed plants that share many morphological traits with flowering plants, although they are not known to have been flowering plants themselves.

In 2013 flowers encased in amber were found and dated 100 million years before present. The amber had frozen the act of sexual reproduction in the process of taking place. Microscopic images showed tubes growing out of pollen and penetrating the flower's stigma. The pollen was sticky, suggesting it was carried by insects.

Recent DNA analysis based on molecular systematics showed that *Amborella trichopoda,* found on the Pacific island of New Caledonia, belongs to a sister group of the other flowering plants, and morphological studies suggest that it has features that may have been characteristic of the earliest flowering plants.

The orders Amborellales, Nymphaeales, and Austrobaileyales diverged as separate lineages from the remaining angiosperm clade at a very early stage in flowering plant evolution.

The great angiosperm radiation, when a great diversity of angiosperms appears in the fossil record, occurred in the mid-Cretaceous (approximately 100 million years ago). However, a study in 2007 estimated that the division of the five most recent (the genus *Ceratophyllum,* the family Chloranthaceae, the eudicots, the magnoliids, and the monocots) of the eight main groups occurred around 140 million years ago. By the late Cretaceous, angiosperms appear to have dominated environments formerly occupied by ferns and cycadophytes, but large canopy-forming trees replaced conifers as the dominant trees only close to the end of the Cretaceous 66 million years ago or even later, at the beginning of the Tertiary. The radiation of herbaceous angiosperms occurred much later. Yet, many fossil plants recognizable as belonging to modern families (including beech, oak, maple, and magnolia) had already appeared by the late Cretaceous.

It is generally assumed that the function of flowers, from the start, was to involve mobile animals in their reproduction processes. That is, pollen can be scattered even if the flower is not brightly colored or oddly shaped in a way that attracts animals; however, by expending the energy required to create such traits, angiosperms can enlist the aid of animals and, thus, reproduce more efficiently.

Two bees on a flower head of Creeping Thistle, *Cirsium arvense*

Island genetics provides one proposed explanation for the sudden, fully developed appearance of flowering plants. Island genetics is believed to be a common source of speciation in general, especially when it comes to radical adaptations that seem to have required inferior transitional forms. Flowering plants may have evolved in an isolated setting like an island or island chain, where the plants bearing them were able to develop a highly specialized relationship with some specific animal (a wasp, for example). Such a relationship, with a hypothetical wasp carrying pollen from one plant to another much the way fig wasps do today, could result in the development of a high degree of specialization in both the plant(s) and their partners. Note that the wasp example is not incidental; bees, which, it is postulated, evolved specifically due to mutualistic plant relationships, are descended from wasps.

Animals are also involved in the distribution of seeds. Fruit, which is formed by the enlargement of flower parts, is frequently a seed-dispersal tool that attracts animals to eat or otherwise disturb it, incidentally scattering the seeds it contains. Although many such mutualistic relationships remain too fragile to survive competition and to spread widely, flowering proved to be an unusually effective means of reproduction, spreading (whatever its origin) to become the dominant form of land plant life.

Flower ontogeny uses a combination of genes normally responsible for forming new shoots. The most primitive flowers probably had a variable number of flower parts, often separate from (but in contact with) each other. The flowers tended to grow in a spiral pattern, to be bisexual (in plants, this means both male and female parts on the same flower), and to be dominated by the ovary (female part). As flowers evolved, some variations developed parts fused together, with a much more specific number and design, and with either specific sexes per flower or plant or at least "ovary-inferior".

Flower evolution continues to the present day; modern flowers have been so profoundly influenced by humans that some of them cannot be pollinated in nature. Many modern domesticated flower species were formerly simple weeds, which sprouted only when the ground was disturbed. Some of them tended to grow with human crops, perhaps already having symbiotic companion plant relationships with them, and the prettiest did not get plucked because of their beauty, developing a dependence upon and special adaptation to human affection.

A few paleontologists have also proposed that flowering plants, or angiosperms, might have evolved due to interactions with dinosaurs. One of the idea's strongest proponents is Robert T. Bakker. He proposes that herbivorous dinosaurs, with their eating habits, provided a selective pressure on plants, for which adaptations either succeeded in deterring or coping with predation by herbivores.

Classification

There are eight groups of living angiosperms:

- *Amborella*, a single species of shrub from New Caledonia;
- Nymphaeales, about 80 species, water lilies and Hydatellaceae;
- Austrobaileyales, about 100 species of woody plants from various parts of the world;
- Chloranthales, several dozen species of aromatic plants with toothed leaves;
- Magnoliids, about 9,000 species, characterized by trimerous flowers, pollen with one pore, and usually branching-veined leaves—for example magnolias, bay laurel, and black pepper;
- Monocots, about 70,000 species, characterized by trimerous flowers, a single cotyledon, pollen with one pore, and usually parallel-veined leaves—for example grasses, orchids, and palms;
- *Ceratophyllum*, about 6 species of aquatic plants, perhaps most familiar as aquarium plants;
- Eudicots, about 175,000 species, characterized by 4- or 5-merous flowers, pollen with three pores, and usually branching-veined leaves—for example sunflowers, petunia, buttercup, apples, and oaks.

The exact relationship between these eight groups is not yet clear, although there is agreement that the first three groups to diverge from the ancestral angiosperm were Amborellales, Nymphaeales, and Austrobaileyales. The term basal angiosperms refers to these three groups. Among the rest, the relationship between the three broadest of these groups (magnoliids, monocots, and eudicots) remains unclear. Some analyses make the magnoliids the first to diverge, others the monocots. *Ceratophyllum* seems to group with the eudicots rather than with the monocots.

History of Classification

The botanical term "Angiosperm", from the Ancient (seed), was coined in the form Angiospermae by Paul Hermann in 1690, as the name of one of his primary divisions of the plant kingdom. This included flowering plants possessing seeds enclosed in capsules, distinguished from his Gymnospermae, or flowering plants with achenial or schizo-carpic fruits, the whole fruit or each of its pieces being here regarded as a seed and naked. The term and its antonym were maintained by Carl Linnaeus with the same sense, but with restricted application, in the names of the orders of his class Didynamia. Its use with any approach to its modern scope became possible only after 1827, when Robert Brown established the existence of truly naked ovules in the Cycadeae and Coniferae, and applied to them the name Gymnosperms. From that time onward, as long as these

Gymnosperms were, as was usual, reckoned as dicotyledonous flowering plants, the term Angiosperm was used antithetically by botanical writers, with varying scope, as a group-name for other dicotyledonous plants.

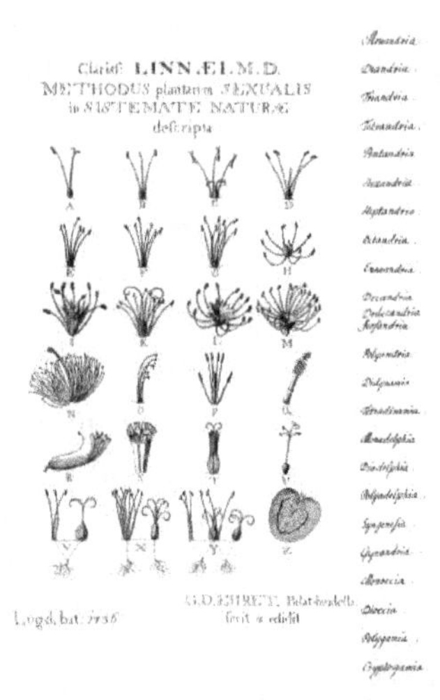

From 1736, an illustration of Linnaean classification.

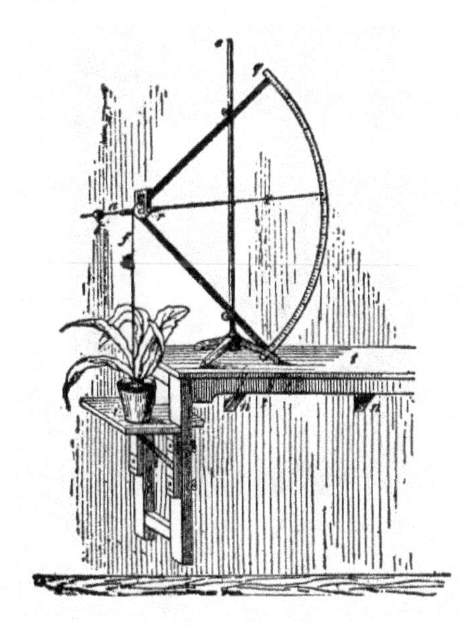

Auxanometer: Device for measuring increase or rate of growth in plants

In 1851, Hofmeister discovered the changes occurring in the embryo-sac of flowering plants, and determined the correct relationships of these to the Cryptogamia. This fixed the position of Gymnosperms as a class distinct from Dicotyledons, and the term Angiosperm then gradually came to be accepted as the suitable designation for the whole of the flowering plants other than Gymnosperms, including the classes of Dicotyledons and Monocotyledons. This is the sense in which the term is used today.

In most taxonomies, the flowering plants are treated as a coherent group. The most popular descriptive name has been Angiospermae (Angiosperms), with Anthophyta ("flowering plants") a second choice. These names are not linked to any rank. The Wettstein system and the Engler system use the name Angiospermae, at the assigned rank of subdivision. The Reveal system treated flowering plants as subdivision Magnoliophytina (Frohne & U. Jensen ex Reveal, Phytologia 79: 70 1996), but later split it to Magnoliopsida, Liliopsida, and Rosopsida. The Takhtajan system and Cronquist system treat this group at the rank of division, leading to the name Magnoliophyta (from the family name Magnoliaceae). The Dahlgren system and Thorne system (1992) treat this group at the rank of class, leading to the name Magnoliopsida. The APG system of 1998, and the later 2003 and 2009 revisions, treat the flowering plants as a clade called angiosperms without a formal botanical name. However, a formal classification was published alongside the 2009 revision in which the flowering plants form the Subclass Magnoliidae.

The internal classification of this group has undergone considerable revision. The Cronquist system, proposed by Arthur Cronquist in 1968 and published in its full form in 1981, is still widely used but is no longer believed to accurately reflect phylogeny. A consensus about how the flowering plants should be arranged has recently begun to emerge through the work of the Angiosperm Phylogeny Group (APG), which published an influential reclassification of the angiosperms in 1998. Updates incorporating more recent research were published as APG II in 2003 and as APG III in 2009.

Monocot (left) and dicot seedlings

Traditionally, the flowering plants are divided into two groups, which in the Cronquist system are called Magnoliopsida (at the rank of class, formed from the family name Magnoliaceae) and Liliopsida (at the rank of class, formed from the family name Liliaceae). Other descriptive names allowed by Article 16 of the ICBN include Dicotyledones or Dicotyledoneae, and Monocotyledones or Monocotyledoneae, which have a long history of use. In English a member of either group may be called a dicotyledon (plural dicotyledons) and monocotyledon (plural monocotyledons), or abbreviated, as dicot (plural dicots) and monocot (plural monocots). These names derive from the observation that the dicots most often have two cotyledons, or embryonic leaves, within each seed. The monocots usually have only one, but the rule is not absolute either way. From a broad diagnostic point of view, the number of cotyledons is neither a particularly handy nor a reliable character.

Recent studies, as by the APG, show that the monocots form a monophyletic group (clade) but that the dicots do not (they are paraphyletic). Nevertheless, the majority of dicot species do form a monophyletic group, called the eudicots or tricolpates. Of the remaining dicot species, most belong to a third major clade known as the magnoliids, containing about 9,000 species. The rest include a paraphyletic grouping of primitive species known collectively as the basal angiosperms, plus the families Ceratophyllaceae and Chloranthaceae.

Flowering Plant Diversity

A poster of twelve different species of flowers of the *Asteraceae* family

Lupinus pilosus

The number of species of flowering plants is estimated to be in the range of 250,000 to 400,000. This compares to around 12,000 species of moss or 11,000 species of pteridophytes, showing that the flowering plants are much more diverse. The number of families in APG (1998) was 462. In APG II (2003) it is not settled; at maximum it is 457, but within this number there are 55 optional segregates, so that the minimum number of families in this system is 402. In APG III (2009) there are 415 families.

The diversity of flowering plants is not evenly distributed. Nearly all species belong to the eudicot (75%), monocot (23%), and magnoliid (2%) clades. The remaining 5 clades contain a little over 250 species in total; i.e. less than 0.1% of flowering plant diversity, divided among 9 families. The 42 most-diverse of 443 families of flowering plants by species, in their APG circumscriptions, are

- Asteraceae or Compositae (daisy family): 22,750 species;

- Orchidaceae (orchid family): 21,950;
- Fabaceae or Leguminosae (bean family): 19,400;
- Rubiaceae (madder family): 13,150;
- Poaceae or Gramineae (grass family): 10,035;
- Lamiaceae or Labiatae (mint family): 7,175;
- Euphorbiaceae (spurge family): 5,735;
- Melastomataceae or Melastomaceae (melastome family): 5,005;
- Myrtaceae (myrtle family): 4,625;
- Apocynaceae (dogbane family): 4,555;
- Cyperaceae (sedge family): 4,350;
- Malvaceae (mallow family): 4,225;
- Araceae (arum family): 4,025;
- Ericaceae (heath family): 3,995;
- Gesneriaceae (gesneriad family): 3,870;
- Apiaceae or Umbelliferae (parsley family): 3,780;
- Brassicaceae or Cruciferae (cabbage family): 3,710:
- Piperaceae (pepper family): 3,600;
- Acanthaceae (acanthus family): 3,500;
- Rosaceae (rose family): 2,830;
- Boraginaceae (borage family): 2,740;
- Urticaceae (nettle family): 2,625;
- Ranunculaceae (buttercup family): 2,525;
- Lauraceae (laurel family): 2,500;
- Solanaceae (nightshade family): 2,460;
- Campanulaceae (bellflower family): 2,380;
- Arecaceae (palm family): 2,361;
- Annonaceae (custard apple family): 2,220;
- Caryophyllaceae (pink family): 2,200;
- Orobanchaceae (broomrape family): 2,060;
- Amaranthaceae (amaranth family): 2,050;

- Iridaceae (iris family): 2,025;
- Aizoaceae or Ficoidaceae (ice plant family): 2,020;
- Rutaceae (rue family): 1,815;
- Phyllanthaceae (phyllanthus family): 1,745;
- Scrophulariaceae (figwort family): 1,700;
- Gentianaceae (gentian family): 1,650;
- Convolvulaceae (bindweed family): 1,600;
- Proteaceae (protea family): 1,600;
- Sapindaceae (soapberry family): 1,580;
- Cactaceae (cactus family): 1,500;
- Araliaceae (*Aralia* or ivy family): 1,450.

Of these, the Orchidaceae, Poaceae, Cyperaceae, Arecaceae, and Iridaceae are monocot families; Piperaceae, Lauraceae, and Annonaceae are magnoliid dicots; the rest of the families are eudicots.

Vascular Anatomy

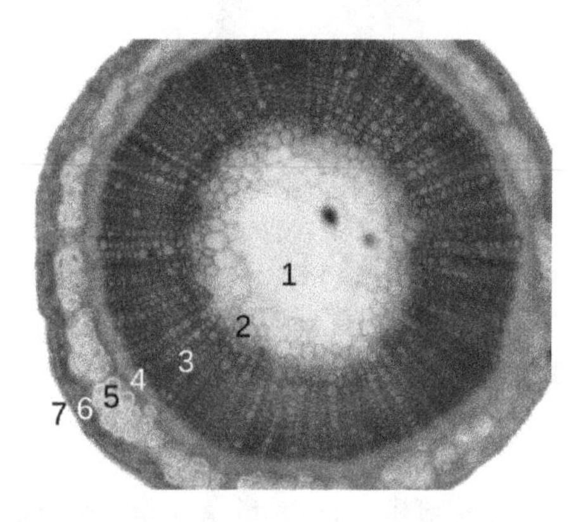

Cross-section of a stem of the angiosperm flax:
1. Pith,
2. Protoxylem,
3. Xylem I,
4. Phloem I,
5. Sclerenchyma (bast fibre),
6. Cortex,
7. Epidermis

The amount and complexity of tissue-formation in flowering plants exceeds that of gymnosperms. The vascular bundles of the stem are arranged such that the xylem and phloem form concentric rings.

In the dicotyledons, the bundles in the very young stem are arranged in an open ring, separating a central pith from an outer cortex. In each bundle, separating the xylem and phloem, is a layer of meristem or active formative tissue known as cambium. By the formation of a layer of cambium between the bundles (interfascicular cambium), a complete ring is formed, and a regular periodical increase in thickness results from the development of xylem on the inside and phloem on the outside. The soft phloem becomes crushed, but the hard wood persists and forms the bulk of the stem and branches of the woody perennial. Owing to differences in the character of the elements produced at the beginning and end of the season, the wood is marked out in transverse section into concentric rings, one for each season of growth, called annual rings.

Among the monocotyledons, the bundles are more numerous in the young stem and are scattered through the ground tissue. They contain no cambium and once formed the stem increases in diameter only in exceptional cases.

The Flower, Fruit, and Seed

Flowers

A collection of flowers forming an inflorescence

The characteristic feature of angiosperms is the flower. Flowers show remarkable variation in form and elaboration, and provide the most trustworthy external characteristics for establishing relationships among angiosperm species. The function of the flower is to ensure fertilization of the ovule and development of fruit containing seeds. The floral apparatus may arise terminally on a shoot or from the axil of a leaf (where the petiole attaches to the stem). Occasionally, as in violets, a flower arises singly in the axil of an ordinary foliage-leaf. More typically, the flower-bearing portion of the plant is sharply distinguished from the foliage-bearing or vegetative portion, and forms a more or less elaborate branch-system called an inflorescence.

There are two kinds of reproductive cells produced by flowers. Microspores, which will divide to become pollen grains, are the "male" cells and are borne in the stamens (or microsporophylls). The "female" cells called megaspores, which will divide to become the egg cell (megagametogenesis), are contained in the ovule and enclosed in the carpel (or megasporophyll).

The flower may consist only of these parts, as in willow, where each flower comprises only a few stamens or two carpels. Usually, other structures are present and serve to protect the sporophylls and to form an envelope attractive to pollinators. The individual members of these surrounding structures are known as sepals and petals (or tepals in flowers such as *Magnolia* where sepals and petals are not distinguishable from each other). The outer series (calyx of sepals) is usually green and leaf-like, and functions to protect the rest of the flower, especially the bud. The inner series (corolla of petals) is, in general, white or brightly colored, and is more delicate in structure. It functions to attract insect or bird pollinators. Attraction is effected by color, scent, and nectar, which may be secreted in some part of the flower. The characteristics that attract pollinators account for the popularity of flowers and flowering plants among humans.

While the majority of flowers are perfect or hermaphrodite (having both pollen and ovule producing parts in the same flower structure), flowering plants have developed numerous morphological and physiological mechanisms to reduce or prevent self-fertilization. Heteromorphic flowers have short carpels and long stamens, or vice versa, so animal pollinators cannot easily transfer pollen to the pistil (receptive part of the carpel). Homomorphic flowers may employ a biochemical (physiological) mechanism called self-incompatibility to discriminate between self and non-self pollen grains. In other species, the male and female parts are morphologically separated, developing on different flowers.

Fertilization and Embryogenesis

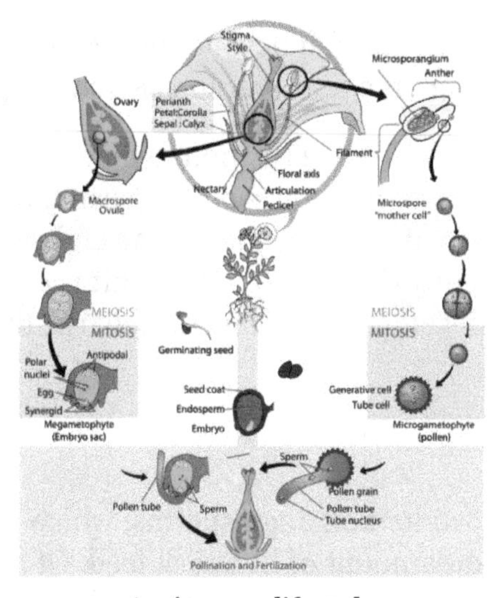

Angiosperm life cycle

Double fertilization refers to a process in which two sperm cells fertilize cells in the ovary. This process begins when a pollen grain adheres to the stigma of the pistil (female reproductive structure), germinates, and grows a long pollen tube. While this pollen tube is growing, a haploid generative cell travels down the tube behind the tube nucleus. The generative cell divides by mitosis to produce two haploid (n) sperm cells. As the pollen tube grows, it makes its way from the stigma, down the style and into the ovary. Here the pollen tube reaches the micropyle of the ovule and digests its way into one of the synergids, releasing its contents (which include the sperm cells). The synergid that the cells were released into degenerates and one sperm makes its way to fertilize the egg cell, producing a diploid ($2n$) zygote. The second sperm cell fuses with both central cell nuclei, producing a triploid ($3n$) cell. As the zygote develops into an embryo, the triploid cell develops into the endosperm, which serves as the embryo's food supply. The ovary will now develop into a fruit and the ovule will develop into a seed.

Fruit and Seed

The fruit of the *Aesculus* or Horse Chestnut tree

As the development of embryo and endosperm proceeds within the embryo sac, the sac wall enlarges and combines with the nucellus (which is likewise enlarging) and the integument to form the *seed coat*. The ovary wall develops to form the fruit or pericarp, whose form is closely associated with the manner of distribution of the seed.

Frequently, the influence of fertilization is felt beyond the ovary, and other parts of the flower take part in the formation of the fruit, e.g., the floral receptacle in the apple, strawberry, and others.

The character of the seed coat bears a definite relation to that of the fruit. They protect the embryo and aid in dissemination; they may also directly promote germination. Among plants with indehiscent fruits, in general, the fruit provides protection for the embryo and secures dissemination. In this case, the seed coat is only slightly developed. If the fruit is dehiscent and the seed is exposed, in general, the seed-coat is well developed, and must discharge the functions otherwise executed by the fruit.

Economic Importance

Agriculture is almost entirely dependent on angiosperms, which provide virtually all plant-based food, and also provide a significant amount of livestock feed. Of all the families of plants, the Poaceae, or grass family (grains), is by far the most important, providing the bulk of all feedstocks (rice, corn – maize, wheat, barley, rye, oats, pearl millet, sugar cane, sorghum). The Fabaceae, or legume family, comes in second place. Also of high importance are the Solanaceae, or nightshade family (potatoes, tomatoes, and peppers, among others), the Cucurbitaceae, or gourd family (also including pumpkins and melons), the Brassicaceae, or mustard plant family (including rapeseed and the innumerable varieties of the cabbage species *Brassica oleracea*), and the Apiaceae, or parsley family. Many of our fruits come from the Rutaceae, or rue family (including oranges, lemons, grapefruits, etc.), and the Rosaceae, or rose family (including apples, pears, cherries, apricots, plums, etc.).

In some parts of the world, certain single species assume paramount importance because of their variety of uses, for example the coconut (*Cocos nucifera*) on Pacific atolls, and the olive (*Olea europaea*) in the Mediterranean region.

Flowering plants also provide economic resources in the form of wood, paper, fiber (cotton, flax, and hemp, among others), medicines (digitalis, camphor), decorative and landscaping plants, and many other uses. The main area in which they are surpassed by other plants — namely, coniferous trees (Pinales), which are non-flowering (gymnosperms) — is timber and paper production.

Monocotyledon

Monocotyledons, commonly referred to as monocots, (Lilianae *sensu* Chase & Reveal) are flowering plants (angiosperms) whose seeds typically contain only one embryonic leaf, or cotyledon. They constitute one of the major groups into which the flowering plants have traditionally been divided, the rest of the flowering plants having two cotyledons and therefore classified as dicotyledons, or dicots. However, molecular phylogenetic research has shown that while the monocots form a monophyletic group or clade (comprising all the descendants of a common ancestor), the dicots do not. Monocots have almost always been recognized as a group, but with various taxonomic ranks and under several different names. The APG III system of 2009 recognises a clade called "monocots" but does not assign it to a taxonomic rank.

The monocots include about 60,000 species. The largest family in this group (and in the flowering plants as a whole) by number of species are the orchids (family Orchidaceae), with more than 20,000 species. About half as many species belong to the true grasses (Poaceae), who are economically the most important family of monocots. In agriculture the majority of the biomass produced comes from monocots. These include not only major grains (rice, wheat, maize, etc.), but also forage grasses, sugar cane, and the bamboos. Other economically important monocot crops include various palms (Arecaceae), bananas (Musaceae), gingers and their relatives, turmeric and cardamom (Zingiberaceae), asparagus and the onions and garlic family (Amaryllidaceae). Many houseplants are monocot epiphytes. Additionally most of the horticultural bulbs, plants cultivated for their blooms, such as lilies, daffodils, irises, amaryllis, cannas, bluebells and tulips, are monocots.

Description

Allium crenulatum, an onion, with a typical monocot perianth and parallel leaf venation

The monocots or monocotyledons have a single cotyledon, or embryonic leaf, in their seeds. Historically, this feature was used to contrast the monocots with the dicotyledons or dicots which typically have two cotyledons; however modern research has shown that the dicots are not a natural group. From a diagnostic point of view the number of cotyledons is neither a particularly useful characteristic (as they are only present for a very short period in a plant's life), nor is it completely reliable.

Comparison with Dicotyledons

Comparison of a monocot (grass) sprouting (left) with a dicot (right), showing hypogeal development in which the cotyledon remains invisible within the seed, underground. The visible part is the first true leaf produced from the meristem; the cotyledon itself remains within the seed.

Slice of onion, showing parallel veins in cross section

Ceroxylon quindiuense (Quindio wax palm) is considered the tallest monocot in the world

The traditionally listed differences between monocotyledons and dicotyledons are as follows. This is a broad sketch only, not invariably applicable, as there are a number of exceptions. The differences indicated are more true for monocots versus eudicots.

Feature	In monocots	In dicots
Growth form	Mostly herbaceous, occasionally arboraceous	Herbaceous or arboraceous
Leaves	Leaf shape oblong or linear, often sheathed at base, petiole seldom developed, stipules absent. Major leaf veins usually parallel	Broad, seldom sheathed, petiole common often with stipules. Veins usually reticulate (pinnate or palmate)
Roots	Primary root of short duration, replaced by adventitial roots forming fibrous or fleshy root systems	Develops from the radicle. Primary root often persists forming strong taproot and secondary roots
Plant stem: Vascular bundles	Numerous scattered bundles in ground parenchyma, cambium rarely present, no differentiation between cortical and stelar regions	Ring of primary bundles with cambium, differentiated into cortex and stele
Flowers	Parts in threes (trimerous) or multiples of three (e.g. 3, 6 or 9 petals)	Fours (tetramerous) or fives (pentamerous)
Pollen: Number of apertures (furrows or pores)	Monocolpate (single aperture or colpus)	Tricolpate (three)
Embryo: Number of cotyledons (leaves in the seed)	One, endosperm frequently present in seed	Two, endosperm present or absent

A number of these differences are not unique to the monocots, and while still useful no one single feature, will infallibly identify a plant as a monocot. For example, trimerous flowers and monosulcate pollen are also found in magnoliids, of which exclusively adventitious roots are found in some of the Piperaceae. Similarly, at least one of these traits, parallel leaf veins, is far from universal among the monocots. Monocots with broad leaves and reticulate leaf veins, typical of dicots, are found in a wide variety of monocot families: for example, *Trillium*, *Smilax* (greenbriar), and *Pogonia* (an orchid), and the Dioscoreales (yams). *Potamogeton* are one of several monocots with tetramerous flowers. Other plants exhibit a mixture of characteristics. Nymphaeaceae (water lilies) have reticulate veins, a single cotyledon, adventitious roots and a monocot like vascular bundle. These examples reflect their shared ancestry. Nevertheless, this list of traits is a generally valid set of contrasts, especially when contrasting monocots with eudicots rather than non-monocot flowering plants in general.

Vascular System

Monocots have a distinctive arrangement of vascular tissue known as an atactostele in which the vascular tissue is scattered rather than arranged in concentric rings. Collenchyma is absent in monocot stems, roots and leaves. Many monocots are herbaceous and do not have the ability to increase the width of a stem (secondary growth) via the same kind of vascular cambium found in non-monocot woody plants. However, some monocots do have secondary growth, and because it does not arise from a single vascular cambium producing xylem inwards and phloem outwards, it is termed “anomalous secondary growth”. Examples of large monocots which either exhibit secondary growth, or can reach large sizes without it, are palms (Arecaceae), screwpines (Pandanaceae), bananas (Musaceae), *Yucca*, *Aloe*, *Dracaena*, and *Cordyline*.

Stems of two *Roystonea regia* palms showing anomalous secondary growth in monocots. Note the characteristic fibrous roots, typical of monocots.

Synapomorphies

By contrast Douglas E. Soltis and others identify thirteen synapomorphies (shared characteristics that unite monophyletic groups of taxa);

- Calcium oxalate raphides
- Absence of vessels in leaves
- Monocotyledonous anther wall formation
- Successive microsporogenesis
- Syncarpous gynoecium
- Parietal placentation
- Monocotyledonous seedling
- Persistent radicle
- Haustorial cotyledon tip
- Open cotyledon sheath
- Steroidal sapanonins
- fly pollination
- diffuse vascular bundles and absence of secondary growth

Taxonomy

The monocots form one of five major lineages of mesangiosperms, which in themselves form 99.95% of all angiosperms. The monocots and the eudicots, are the largest and most diversified angiosperm radiations accounting for 20% and 75% of all angiosperm species respectively.

Monocot diversity includes perennial geophytes including ornamental flowers (orchids, tulips and lilies) (Asparagales, Liliales respectively), rosette and succulent epiphytes (Asparagales), myco-heterotrophs (Liliales, Dioscoreales, Pandanales), all in the lilioid monocots, major grains (maize, rice and wheat) in the grass family (Poales) as well as woody tree-like palm trees (Arecales) and bamboo (Poales) in the commelinid monocots, as well as both emergent (Poales, Acorales) and floating or submerged aquatic plants (Alismatales).

Early History

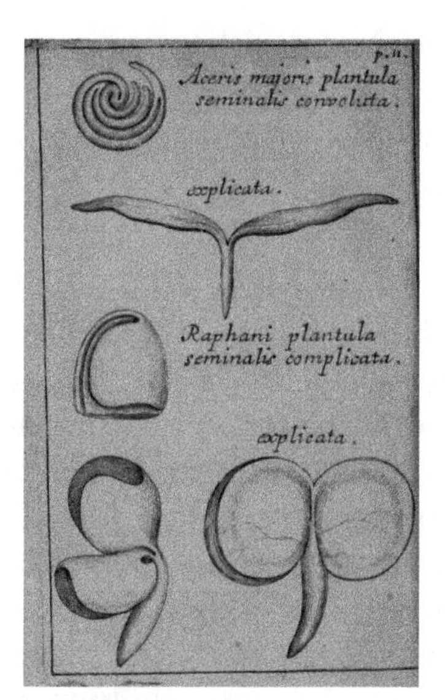

Illustrations of cotyledons by John Ray 1682, after Malpighi

The monocots are one of the major divisions of the flowering plants or angiosperms. They have been recognized as a natural group since John Ray's studies of seed structure in the 17th century. Ray was the first botanical systematist, and in his examination of seeds, first observed the dichotomy of cotyledon structure. He reported his findings in a paper read to the Royal Society on 17 December 1674, entitled "A Discourse on the Seeds of Plants".

A Discourse on the Seeds of Plants

The greatest number of plants that come of seed spring at first out of the earth with two leaves which being for the most part of a different figure from the succeeding leaves are by our gardeners not improperly called the seed leaves...
In the first kind the seed leaves are nothing but the two lobes of the seed having their plain sides clapt together like the two halfs of a walnut and therefore are of the just figure of the seed slit in sunder flat wise...
Of seeds that spring out of the earth with leaves like the succeeding and no seed leaves I have observed two sorts. 1. Such as are congenerous to the first kind precedent that is whose pulp is divided into two lobes and a radicle...
2. Such which neither spring out of the ground with seed leaves nor have their pulp divided into lobes

John Ray (1674), pp. 164, 166

Since this paper appeared a year before the publication of Malpighi's *Anatome Plantarum* (1675–

1679), Ray has the priority. At the time, Ray did not fully realise the importance of his discovery but progressively developed this over successive publications. And since these were in Latin, "seed leaves" became *folia seminalia* and then *cotyledon*, following Malpighi. Malpighi and Ray were familiar with each other's work, and Malpighi in describing the same structures had introduced the term cotyledon, which Ray adopted in his subsequent writing.

De seminum vegetatione

Mense quoque Maii, alias seminales plantulas Fabarum, & Phaseolorum, ablatis pariter binis seminalibus foliis, seu cotyledonibus, incubandas posui
In the month of May, also, I incubated two seed plants, Faba and Phaseolus, after removing the two seed leaves, or cotyledons

Marcello Malpighi (1679), p. 18

In this experiment, Malpighi also showed that the cotyledons were critical to the development of the plant, proof that Ray required for his theory. In his *Methodus plantarum nova* Ray also developed and justified the "natural" or pre-evolutionary approach to classification, based on characteristics selected *a posteriori* in order to group together taxa that have the greatest number of shared characteristics. This approach, also referred to as polythetic would last till evolutionary theory enabled Eichler to develop the phyletic system that superseded it in the late nineteenth century, based on an understanding of the acquisition of characteristics. He also made the crucial observation *Ex hac seminum divisione sumum potest generalis plantarum distinctio, eaque meo judicio omnium prima et longe optima, in eas sci. quae plantula seminali sunt bifolia aut διλόβω, et quae plantula sem. adulta analoga.* (From this division of the seeds derives a general distinction amongst plants, that in my judgement is first and by far the best, into those seed plants which are bifoliate, or bilobed, and those that are analogous to the adult), that is between monocots and dicots. He illustrated this with by quoting from Malpighi and including reproductions of Malpighi's drawings of cotyledons. Initially Ray did not develop a classification of flowering plants (florifera) based on a division by the number of cotyledons, but developed his ideas over successive publications, coining the terms *Monocotyledones* and *Dicotyledones* in 1703, in the revised version of his *Methodus* (*Methodus plantarum emendata*), as a primary method for dividing them, *Herbae floriferae, dividi possunt, ut diximus, in Monocotyledones & Dicotyledones* (Flowering plants, can be divided, as we have said, into Monocotyledons & Dicotyledons).

Although Linnaeus did not utilise Ray's discovery, basing his own classification solely on floral reproductive morphology, every taxonomist since then, starting with De Jussie and De Candolle, has used Ray's distinction as a major classification characteristic.

Modern Era

Modern research based on DNA has confirmed the status of the monocots as a monophyletic group or clade, in contrast to the other historical divisions of the flowering plants, which have had to be substantially reorganized. The monocots form about a quarter of all of the Angiosperms (flowering plants). Of some 60,000 species, by far the largest number (65%) are found in two families, the orchids and grasses. The orchids (Orchidaceae, Asparagales) contain about 25,000 species and the grasses (Poaceae, Poales) about 11,000. Other well known groups within the Poales order include the Cyperaceae (sedges) and Juncaceae (rushes), and the monocots also include familiar families such as the palms (Arecaceae, Arecales) and lilies (Liliaceae, Liliales).

Taxonomists had considerable latitude in naming this group, as the monocots are a group above the rank of family. Article 16 of the *ICBN* allows either a descriptive name or a name formed from the name of an included family.

Historically, the monocotyledons were named:

- Monocotyledoneae in the de Candolle system and the Engler system
- Monocotyledones in the Bentham & Hooker system and the Wettstein system
- class Liliopsida in the Takhtajan system and the Cronquist system
- subclass Liliidae in the Dahlgren system and the Thorne system (1992)
- clade monocots in the Angiosperm Phylogeny Group (APG) systems: the APG system, the APG II system, the APG III system and the APG IV system

Until the rise of the phylogenetic APG systems, it was widely accepted that angiosperms were neatly split between monocots and dicots, a state reflected in virtually all the systems. It is now understood that various groups, notably the Magnoliids and ancient lineages known as the basal angiosperms fall outside of this dichotomy. Each of these systems uses its own internal taxonomy for the group. The monocotyledons are famous as a group that is extremely stable in its outer borders (it is a well-defined, coherent group), while in its internal taxonomy is extremely unstable (historically no two authoritative systems have agreed with each other on how the monocotyledons are related to each other).

Molecular studies have both confirmed the monophyly of the monocots and helped elucidate relationships within this group. The APG III system does not assign the monocots to a taxonomic rank, instead recognizing a monocots clade. However, there has remained some uncertainty regarding the exact relationships between the major lineages, with a number of competing models (including APG).

Subdivisions

Historically, Bentham (1877), considered the monocots to consist of four alliances, Epigynae, Coronariae, Nudiflorae and Glumales, based on floral characteristics. He describes the attempts to subdivide the group since the days of Lindley as largely unsuccessful. Like most subsequent classification systems it failed to distinguish between two major orders, Liliales and Asparagales, now recognised as quite separate. A major advance in this respect was the work of Rolf Dahlgren (1980), which would form the basis of the Angiosperm Phylogeny Group's (APG) subsequent modern classification of monocot families. Dahlgren who used the alternate name Lilliidae considered the monocots as a subclass of angiosperms characterised by a single cotyledon and the presence of triangular protein bodies in the sieve tube plastids. He divided the monocots into seven superorders, Alismatiflorae, Ariflorae, Triuridiflorae, Liliiflorae, Zingiberiflorae, Commeliniflorae and Areciflorae. With respect to the specific issue regarding Liliales and Asparagales, Dahlgren followed Huber (1969) in adopting a splitter approach, in contrast to the longstanding tendency to view Liliaceae as a very broad sensu lato family. Following Dahlgren's untimely death in 1987, his work was continued by his widow, Gertrud Dahlgren, who published a revised version of the classification in 1989. In this scheme the suffix *-florae* was replaced with *-anae* (*e.g.* Alismatanae) and the number of superorders expanded to ten with the addition of Bromelianae, Cyclanthanae and Pandananae.

The APG system establishes ten orders of monocots and two families of monocots (Petrosaviaceae and Dasypogonaceae) not yet assigned to any order. More recently, the Petrosaviaceae has been included in the Petrosaviales, and placed near the lilioid orders. The family Hydatellaceae, assigned to order Poales in the APG II system, has since been recognized as being misplaced in the monocots, and instead proves to be most closely related to the water lilies, family Nymphaeaceae. Family Dasypogonaceae is placed in order Arecales in the APG IV system.

Evolution

The monocots form a monophyletic group arising early in the history of the flowering plants, but the fossil record is meagre. The earliest fossils presumed to be monocot remains date from the early Cretaceous period. For a very long time, fossils of palm trees were believed to be the oldest monocots, first appearing 90 million years ago, but this estimate may not be entirely true. At least some putative monocot fossils have been found in strata as old as the eudicots. The oldest fossils that are unequivocally monocots are pollen from the Late Barremian–Aptian – Early Cretaceous period, about 120-110 million years ago, and are assignable to clade-Pothoideae-Monstereae Araceae; being Araceae, sister to other Alismatales. They have also found flower fossils of Triuridaceae (Pandanales) in Upper Cretaceous rocks in New Jersey, becoming the oldest known sighting of saprophytic/mycotrophic habits in angiosperm plants and among the oldest known fossils of monocotyledons.

Topology of the angiosperm phylogenetic tree could infer that the monocots would be among the oldest lineages of angiosperms, which would support the theory that they are just as old as the eudicots. The pollen of the eudicots dates back 125 million years, so the lineage of monocots should be that old too.

Molecular Clock Estimates

Kåre Bremer, using rbcL sequences and the mean path length method ("mean-path lengths method"), estimated the age of the monocot crown group (i.e. the time at which the ancestor of today's *Acorus* diverged from the rest of the group) as 134 million years. Similarly, Wikström *et al.*, using Sanderson's non-parametric rate smoothing approach ("nonparametric rate smoothing approach"), obtained ages of 158 or 141 million years for the crown group of monocots. All these estimates have large error ranges (usually 15-20%), and Wikström *et al.* used only a single calibration point, namely the split between Fagales and Cucurbitales, which was set to 84 Ma, in the late Santonian period). Early molecular clock studies using strict clock models had estimated the monocot crown age to 200 ± 20 million years ago or 160 ± 16 million years, while studies using relaxed clocks have obtained 135-131 million years or 133.8 to 124 million years. Bremer's estimate of 134 million years has been used as a secondary calibration point in other analyses. Some estimates place the emergence of the monocots as far back as 150 mya in the Jurassic period.

Core Group

The age of the core group of so-called 'nuclear monocots' or 'core monocots', which correspond to all orders except Acorales and Alismatales, is about 131 million years to present, and crown group age is about 126 million years to the present. The subsequent branching in this part of the tree (i.e. Petrosaviaceae, Dioscoreales + Pandanales and Liliales clades appeared), including the crown Petrosaviaceae group may be in the period around 125–120 million years BC (about 111 million years so far), and stem groups of all other orders, including Commelinidae would have diverged about or shortly after 115 million years. These and many clades within these orders may have originated in southern Gondwana, i.e. Antarctica, Australasia, and southern South America.

Aquatic Monocots

The aquatic monocots of Alismatales have commonly been regarded as "primitive". They have also been considered to have the most primitive foliage, which were cross-linked as Dioscoreales and Melanthiales. Keep in mind that the "most primitive" monocot is not necessarily "the sister of everyone else". This is because the ancestral or primitive characters are inferred by means of the reconstruction of character states, with the help of the phylogenetic tree. So primitive characters of monocots may be present in some derived groups. On the other hand, the basal taxa may exhibit many morphological autapomorphies. So although Acoraceae is the sister group to the remaining monocotyledons, the result does not imply that Acoraceae is "the most primitive monocot" in terms of its character states. In fact, Acoraceae is highly derived in many morphological characters, and that is precisely why Acoraceae and Alismatales occupied relatively derived positions in the trees produced by Chase *et al.* and others.

Some authors support the idea of an aquatic phase as the origin of monocots. The phylogenetic position of Alismatales (many water), which occupy a relationship with the rest except the Acoraceae, do not rule out the idea, because it could be 'the most primitive monocots' but not 'the most basal'. The Atactostele stem, the long and linear leaves, the absence of secondary growth, roots in groups instead of a single root branching (related to the nature of the substrate), including sympodial use, are consistent with a water source. However, while monocots were sisters of the aquatic Ceratophyllales, or their origin is related to the adoption of some form of aquatic habit, it would not help much to the understanding of how it evolved to develop their distinctive anatomical features: the monocots seem so different from the rest of angiosperms and it's difficult to relate their morphology, anatomy and development and those of broad-leaved angiosperms.

Other Taxa

In the past, taxa which had petiolate leaves with reticulate venation were considered "primitive" within the monocots, because of its superficial resemblance to the leaves of dicotyledons. Recent work suggests that these taxa are sparse in the phylogenetic tree of monocots, such as fleshy fruited taxa (excluding taxa with aril seeds dispersed by ants), the two features would be adapted to conditions that evolved together regardless. Among the taxa involved were *Smilax*, *Trillium* (Liliales), *Dioscorea* (Dioscoreales), etc. A number of these plants are vines that tend to live in shaded habitats for at least part of their lives, and may also have a relationship with their shapeless stomata. Reticulate venation seems to have appeared at least 26 times in monocots, in fleshy fruits 21 times (sometimes lost later), and the two characteristics, though different, showed strong signs of a tendency to be good or bad in tandem, a phenomenon described as "concerted convergence" ("coordinated convergence").

Etymology

The name monocotyledons is derived from the traditional botanical name "Monocotyledones", which refers to the fact that most members of this group have one cotyledon, or embryonic leaf, in their seeds.

Ecology

Emergence

Some monocots, such as grasses, have hypogeal emergence, where the mesocotyl elongates and pushes the coleoptile (which encloses and protects the shoot tip) toward the soil surface. Since elongation occurs above the cotyledon, it is left in place in the soil where it was planted. Many dicots have epigeal emergence, in which the hypocotyl elongates and becomes arched in the soil. As the hypocotyl continues to elongate, it pulls the cotyledons upward, above the soil surface.

Uses

Of the monocots, the grasses are of enormous economic importance as a source of animal and human food, and form the largest component of agricultural species in terms of biomass produced.

Acorus

Acorus is a genus of monocot flowering plants. This genus was once placed within the family Araceae (aroids), but more recent classifications place it in its own family Acoraceae and order Acorales, of which it is the sole genus of the oldest surviving line of monocots. Some older studies indicated that was placed in a lineage (the order Alismatales), that also includes aroids (Araceae), Tofieldiaceae, and several families of aquatic monocots (e.g., Alismataceae, Posidoniaceae). However, modern phylogenetic studies demonstrate that *Acorus* is sister to all other monocots. Common names include Calamus and Sweet Flag.

The genus is native to North America and northern and eastern Asia, and naturalised in southern Asia and Europe from ancient cultivation. The known wild populations are diploid except for some tetraploids in eastern Asia, while the cultivated plants are sterile triploids, probably of hybrid origin between the diploid and tetraploid forms.

Characteristics

The inconspicuous flowers are arranged on a lateral spadix (a thickened, fleshy axis). Unlike aroids, there is no spathe (large bract, enclosing the spadix). The spadix is 4–10 cm long and is enclosed by the foliage. The bract can be ten times longer than the spadix. The leaves are linear with entire margin.

Habit of *Acorus calamus*.

Taxonomy

Although the family Acoraceae was originally described in 1820, since then *Acorus* has traditionally been included in Araceae in most classification systems, as in the Cronquist system. The family has recently been resurrected as molecular systematic studies have shown that *Acorus* is not closely related to Araceae or any other monocot family, leading plant systematists to place the genus and family in its own order. This placement currently lacks support from traditional plant morphology studies, and some taxonomists still place it as a subfamily of Araceae, in the order Alismatales. The APG III system recognizes order Acorales, distinct from the Alismatales, and as the sister group to all other monocots. This relationship is confirmed by more recent phylogenetic studies.

Species

In older literature and on many websites, there is still much confusion, with the name *Acorus calamus* equally but wrongfully applied to *Acorus americanus* (formerly *Acorus calamus* var. *americanus*).

As of July 2014, the Kew Checklist accepts only 2 species, one of which has three accepted varieties:

- *Acorus calamus* L. – Common Sweet Flag; sterile triploid ($3n = 36$); probably of cultivated origin. It is native to Europe, temperate India and the Himalayas and southern Asia, widely cultivated and naturalised elsewhere.
    - *Acorus calamus* var. *americanus* Raf. - Canada, northern United States, Buryatiya region of Russia
    - *Acorus calamus* var. *angustatus* Besser - Siberia, China, Russian Far East, Japan, Korea, Mongolia, Himalayas, Indian Subcontinent, Indochina, Philippines, Indonesia
    - *Acorus calamus* var. *calamus* - Siberia, Russian Far east, Mongolia, Manchuria, Korea, Himalayas; naturalized in Europe, North America, Java and New Guinea

- *Acorus gramineus* Sol. ex Aiton – Japanese Sweet Flag or Grassy-leaved Sweet Flag; fertile diploid ($2n = 18$); - China, Himalayas, Japan, Korea, Indochina, Philippines, Primorye

Acorus from Europe, China and Japan have been planted in the United States.

Etymology

The name 'acorus' is derived from the Greek word 'acoron', a name used by Dioscorides, which in turn was derived from 'coreon', meaning 'pupil', because it was used in herbal medicine as a treatment for inflammation of the eye.

Distribution and Habitat

These plants are found in wetlands, particularly marshes, where they spread by means of thick rhizomes. Like many other marsh plants, they depend upon aerenchyma to transport oxygen to the rooting zone. They frequently occur on shorelines and floodplains where water levels fluctuate seasonally.

Ecology

The native North American species appears in many ecological studies. Compared to other species of wetland plants, they have relatively high competitive ability. Although many marsh plants accumulate large banks of buried seeds, seed banks of *Acorus* may not accumulate in some wetlands owing to low seed production. The seeds appear to be adapted to germinate in clearings; after a period of cold storage, the seeds will germinate after seven days of light with fluctuating temperature, and somewhat longer under constant temperature. A comparative study of its life history traits classified it as a "tussock interstitial", that is, a species that has a dense growth form and tends to occupy gaps in marsh vegetation, not unlike *Iris versicolor*.

Toxicity

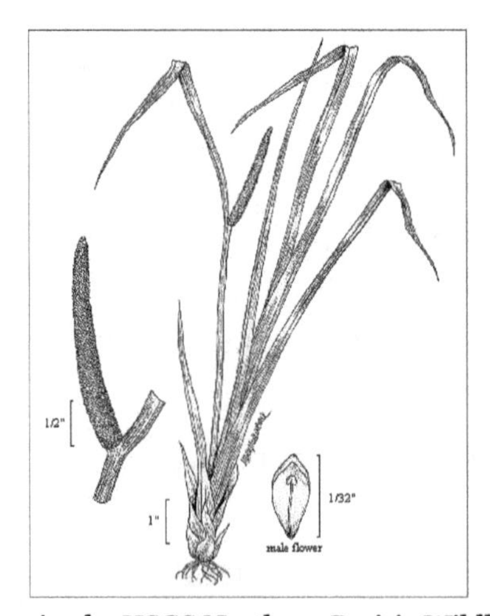

Sweet Flag (2006 drawing by USGS Northern Prairie Wildlife Research Center)

Products derived from *Acorus calamus* were banned in 1968 as food additives by the United States Food and Drug Administration. The questionable chemical derived from the plant was β-asarone. Confusion exists whether all strains of *A. calamus* contain this substance.

Four varieties of *A. calamus* strains exist in nature: diploid, triploid, tetraploid and hexaploid. Diploids do not produce the carcinogenic β-asarone. Diploids are known to grow naturally in Eastern Asia (Mongolia and C Siberia) and North America. The triploid cytotype probably originated in the Himalayan region, as a hybrid between the diploid and tetraploid cytotypes. The North American Calamus is known as *Acorus calamus* var. *americanus* or more recently as simply *Acorus americanus*. Like the diploid strains of *A. calamus* in parts of the Himalayas, Mongolia, and C Siberia, the North American diploid strain does not contain the carcinogenic β-asarone. Research has consistently demonstrated that "β-asarone was not detectable in the North American spontaneous diploid Acorus [Calamus var. Americanus]".

Uses

The parallel-veined leaves of some species contain ethereal oils that give a sweet scent when dried. Fine-cut leaves used to be strewn across the floor in the Middle Ages, both for the scent, and for presumed efficacy against pests.

References

- Taylor, T.N.; Taylor, E.L. & Krings, M. (2009), Paleobotany, The Biology and Evolution of Fossil Plants (2nd ed.), Amsterdam; Boston: Academic Press, ISBN 978-0-12-373972-8, p. 1027
- Kenrick, P. & Crane, P.R. (1997), The Origin and Early Diversification of Land Plants: A Cladistic Study, Washington, D.C.: Smithsonian Institution Press, ISBN 978-1-56098-730-7
- Novíkov & Barabaš-Krasni (2015). "Modern plant systematics". Liga-Pres: 685. doi:10.13140/RG.2.1.4745.6164. ISBN 978-966-397-276-3.
- Anderson, Anderson & Cleal (2007). "Brief history of the gymnosperms: classification, biodiversity, phytogeography and ecology". Strelitzia. SANBI. 20: 280. ISBN 978-1-919976-39-6.
- Taylor, T.N.; Taylor, E.L. & Krings, M. (2009), Paleobotany, The Biology and Evolution of Fossil Plants (2nd ed.), Amsterdam; Boston: Academic Press, ISBN 978-0-12-373972-8, pp. 508ff.
- Novíkov & Barabaš-Krasni (2015). "Modern plant systematics". Liga-Pres: 685. doi:10.13140/RG.2.1.4745.6164. ISBN 978-966-397-276-3.
- Anderson, Anderson & Cleal (2007). "Brief history of the gymnosperms: classification, biodiversity, phytogeography and ecology". Strelitzia. SANBI. 20: 280. ISBN 978-1-919976-39-6.
- Kotpal, Tyagi, Bendre, & Pande. Concepts of Biology XI. Rastogi Publications, 2nd ed. New Delhi 2007. ISBN 8171338968. Fig. 38 Types of placentation, page 2-127
- Stewart, W.N.; Rothwell, G.W. (1993). Paleobotany and the evolution of plants. Cambridge University Press. ISBN 0521382947.
- Walters, Dirk R Walters Bonnie By (1996). Vascular plant taxonomy. Dubuque, Iowa: Kendall/Hunt Pub. Co. p. 124. ISBN 978-0-7872-2108-9.
- David Sadava; H. Craig Heller; Gordon H. Orians; William K. Purves; David M. Hillis (December 2006). Life: the science of biology. Macmillan. pp. 477–. ISBN 978-0-7167-7674-1. Retrieved 4 August 2010.
- Stewart, Wilson Nichols; Rothwell, Gar W. (1993). Paleobotany and the evolution of plants (2nd ed.). Cambridge Univ. Press. p. 498. ISBN 0-521-23315-1.

Pollination: A Comprehensive Study

Pollination is the process in which pollen is transferred to the female reproductive organ. This process enables fertilization to take place. The forms of pollination elucidated within the section are anemophily, hydrophily, entomophily and zoophily. The chapter strategically encompasses and incorporates the major components of pollination, providing a complete understanding.

Pollination

Carpenter bee with pollen collected from Night-blooming cereus

Pollination is the process by which pollen is transferred to the female reproductive organs of a plant, thereby enabling fertilization to take place. Like all living organisms, seed plants have a single major goal: to pass their genetic information on to the next generation. The reproductive unit is the seed, and pollination is an essential step in the production of seeds in all spermatophytes (seed plants).

For the process of pollination to be successful, a pollen grain produced by the anther, the male part of a flower, must be transferred to a stigma, the female part of the flower, of a plant of the same species. The process is rather different in angiosperms (flowering plants) from what it is in gymnosperms (other seed plants). In angiosperms, after the pollen grain has landed on the stigma, it creates a pollen tube which grows down the style until it reaches the ovary. Sperm cells from the pollen grain then move along the pollen tube, enter the egg cell through the micropyle and fertilise it, resulting in the production of a seed.

A successful angiosperm pollen grain (gametophyte) containing the male gametes is transported to the stigma, where it germinates and its pollen tube grows down the style to the ovary. Its two gametes travel down the tube to where the gametophyte(s) containing the female gametes are held within the carpel. One nucleus fuses with the polar bodies to produce the endosperm tissues, and the other with the ovule to produce the embryo Hence the term: "double fertilization".

Tip of a tulip stamen covered with pollen grains.

In gymnosperms, the ovule is not contained in a carpel, but exposed on the surface of a dedicated support organ, such as the scale of a cone, so that the penetration of carpel tissue is unnecessary. Details of the process vary according to the division of gymnosperms in question. Two main modes of fertilization are found in gymnosperms. Cycads and *Ginkgo* have motile sperm that swim directly to the egg inside the ovule, whereas conifers and gnetophytes have sperm that are unable to swim but are conveyed to the egg along a pollen tube.

The study of pollination brings together many disciplines, such as botany, horticulture, entomology, and ecology. The pollination process as an interaction between flower and pollen vector was first addressed in the 18th century by Christian Konrad Sprengel. It is important in horticulture and agriculture, because fruiting is dependent on fertilization: the result of pollination. The study of pollination by insects is known as *anthecology*.

Pollination Process

Pollen germination has three stages; hydration, activation and pollen tube emergence. The pollen grain is severely dehydrated so that its mass is reduced enabling it to be more easily transported from flower to flower. Germination only takes place after rehydration, ensuring that premature germination does not take place in the anther. Hydration allows the plasma membrane of the pollen grain to reform into its normal bilayer organization providing an effective osmotic membrane. Activation involves the development of actin filaments throughout the cytoplasm of the cell, which eventually become concentrated at the point from which the pollen tube will emerge. Hydration and activation continue as the pollen tube begins to grow.

In conifers, the reproductive structures are borne on cones. The cones are either pollen cones (male) or ovulate cones (female), but some species are monoecious and others dioecious. A pollen cone contains hundreds of microsporangia carried on (or borne on) reproductive structures called sporophylls. Spore mother cells in the microsporangia divide by meiosis to form haploid microspores that develop further by two mitotic divisions into immature male gametophytes (pollen grains). The four resulting cells consist of a large tube cell that forms the pollen tube, a generative cell that will produce two sperm by mitosis, and two prothallial cells that degenerate. These cells comprise a very reduced microgametophyte, that is contained within the resistant wall of the pollen grain.

The pollen grains are dispersed by the wind to the female, ovulate cone that is made up of many overlapping scales (sporophylls, and thus megasporophylls), each protecting two ovules, each of which consists of a megasporangium (the nucellus) wrapped in two layers of tissue, the integument and the cupule, that were derived from highly modified branches of ancestral gymnosperms. When a pollen grain lands close enough to the tip of an ovule, it is drawn in through the micropyle (a pore in the integuments covering the tip of the ovule) often by means of a drop of liquid known as a pollination drop. The pollen enters a pollen chamber close to the nucellus, and there it may wait for a year before it germinates and forms a pollen tube that grows through the wall of the megasporangium (=nucellus) where fertilisation takes place. During this time, the megaspore mother cell divides by meiosis to form four haploid cells, three of which degenerate. The surviving one develops as a megaspore and divides repeatedly to form an immature female gametophyte (egg sac). Two or three archegonia containing an egg then develop inside the gametophyte. Meanwhile, in the spring of the second year two sperm cells are produced by mitosis of the body cell of the male gametophyte. The pollen tube elongates and pierces and grows through the megasporangium wall and delivers the sperm cells to the female gametophyte inside. Fertilisation takes place when the nucleus of one of the sperm cells enters the egg cell in the megagametophyte's archegonium.

In flowering plants, the anthers of the flower produce microspores by meiosis. These undergo mitosis to form male gametophytes, each of which contains two haploid cells. Meanwhile, the ovules produce megaspores by meiosis, further division of these form the female gametophytes, which are very strongly reduced, each consisting only of a few cells, one of which is the egg. When a pollen grain adheres to the stigma of a carpel it germinates, developing a pollen tube that grows through the tissues of the style, entering the ovule through the micropyle. When the tube reaches the egg sac, two sperm cells pass through it into the female gametophyte and fertilisation takes place.

Types

Abiotic

Abiotic pollination refers to situations where pollination is mediated without the involvement of other organisms. The most common form of abiotic pollination, anemophily, is pollination by wind. Wind pollination is very imprecise, with a minute proportion of pollen grains landing by chance on a suitable receptive stigma, the rest being wasted in the environment. This form of pollination is used by grasses, most conifers, and many deciduous trees. Hydrophily is pollination by water, and occurs in aquatic plants which release their pollen directly into the surrounding water. About 80% of all plant pollination is biotic. In gymnosperms, biotic pollination is generally incidental when it occurs, though some gymnosperms and their pollinators are mutually adapted for pollination. The best-known examples probably are members of the order Cycadales and associated species of beetles. Of the abiotically pollinated species of plant, 98% are anemophilous and 2% hydrophilous, their pollen being transported by water.

It is thought that among angiosperms, entomophily is the primitive state; this is indicated by the vestigial nectaries in the wind-pollinated *Urtica* and other plants, and the presence of fragrances in some of these plants. Of the angiosperms, grasses, sedges, rushes and catkin-bearing plants are in general wind pollinated. Other flowering plants are mostly biotic, the pollen being carried by animal vectors. However a number of plants in multiple families have secondarily adopted wind pollination in contrast to other members of their groups. Some plants are intermediate between the two pollination methods. common heather is regularly pollinated by insects, but produce clouds of pollen and some wind pollination is inevitable, and the hoary plantain is primarily wind pollinated, but is also visited by insects which pollinate it.

Melissodes desponsa covered in pollen

Biotic

Hummingbirds typically feed on red flowers

More commonly, the process of pollination requires pollinators: organisms that carry or move the pollen grains from the anther of one flower to the receptive part of the carpel or pistil (stigma) of another. This is biotic pollination. The various flower traits (and combinations thereof) that differentially attract one type of pollinator or another are known as pollination syndromes. At least 100,000 species of animal, and possibly as many as 200,000, act as pollinators of the estimated 250,000 species of flowering plants in the world. The majority of these pollinators are insects, but about 1,500 species of birds and mammals have been reported to visit flowers and may transfer pollen between them. Besides birds and bats which are the most frequent visitors, these include monkeys, lemurs, squirrels, rodents and possums.

Entomophily, pollination by insects, often occurs on plants that have developed colored petals and a strong scent to attract insects such as, bees, wasps and occasionally ants (Hymenoptera), beetles (Coleoptera), moths and butterflies (Lepidoptera), and flies (Diptera). The existence of insect pollination dates back to the dinosaur era.

In zoophily, pollination is performed by vertebrates such as birds and bats, particularly, hummingbirds, sunbirds, spiderhunters, honeyeaters, and fruit bats. Plants adapted to using bats or moths as pollinators typically have white petals and a strong scent and flower at night, whereas plants that use birds as pollinators tend to produce copious nectar and have red petals.

Insect pollinators such as honeybees (*Apis mellifera*), bumblebees (*Bombus terrestris*), and butterflies (*Thymelicus flavus*) have been observed to engage in flower constancy, which means they are more likely to transfer pollen to other conspecific plants. This can be beneficial for the pollinators, as flower constancy prevents the loss of pollen during interspecific flights and pollinators from clogging stigmas with pollen of other flower species. It also improves the probability that the pollinator will find productive flowers easily accessible and recognisable by familiar clues.

Mechanism

A European honey bee collects nectar, while pollen collects on its body.

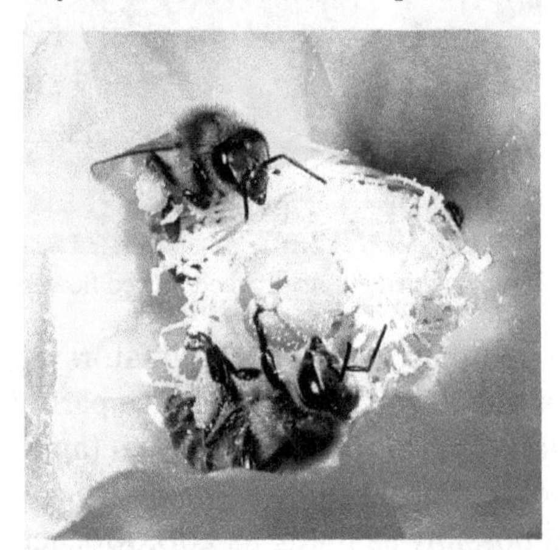

Africanized honey bees immersed in Yellow Opuntia engelmannii Cactus Flower Pollen

Pollination can be accomplished by cross-pollination or by self-pollination:

- Cross-pollination, also called *allogamy*, occurs when pollen is delivered from the stamen of one flower to the stigma of a flower on another plant of the same species. Plants adapted for cross-pollination have several mechanisms to prevent self-pollination; the reproductive organs may be arranged in such a way that self-fertilisation is unlikely, or the stamens and carpels may mature at different times.
- Self-pollination occurs when pollen from one flower pollinates the same flower or other flowers of the same individual. It is thought to have evolved under conditions when pollinators were not reliable vectors for pollen transport, and is most often seen in short-lived

annual species and plants that colonize new locations. Self-pollination may include *autogamy*, where pollen is transferred to the female part of the same flower; or *geitonogamy*, when pollen is transferred to another flower on the same plant. Plants adapted to self-fertilize often have similar stamen and carpel lengths. Plants that can pollinate themselves and produce viable offspring are called self-fertile. Plants that cannot fertilize themselves are called self-sterile, a condition which mandates cross-pollination for the production of offspring.

- *Cleistogamy*: is self-pollination that occurs before the flower opens. The pollen is released from the anther within the flower or the pollen on the anther grows a tube down the style to the ovules. It is a type of sexual breeding, in contrast to asexual systems such as apomixis. Some *cleistogamous* flowers never open, in contrast to *chasmogamous* flowers that open and are then pollinated. Cleistogamous flowers are by necessity found on self-compatible or self-fertile plants. Although certain orchids and grasses are entirely cleistogamous, other plants resort to this strategy under adverse conditions. Often there may be a mixture of both cleistogamous and chasmogamous flowers, sometimes on different parts of the plant and sometimes in mixed inflorescences. The ground bean produces cleistogamous flowers below ground, and mixed cleistogamous and chasmogamous flowers above.

Geranium incanum, like most geraniums and pelargoniums, sheds its anthers, sometimes its stamens as well, as a barrier to self-pollination. This young flower is about to open its anthers, but has not yet fully developed its pistil.

These *Geranium incanum* flowers have opened their anthers, but not yet their stigmas. Note the change of colour that signals to pollinators that it is ready for visits.

An estimated 48.7% of plant species are either dioecious or self-incompatible obligate out-crossers. It is also estimated that about 42% of flowering plants exhibit a mixed mating system in nature. In the most common kind of mixed mating system, individual plants produce a single type of flower and fruits may contain self-pollinated, out-crossed or a mixture of progeny types.

This *Geranium incanum* flower has shed its stamens, and deployed the tips of its pistil without accepting pollen from its own anthers. (It might of course still receive pollen from younger flowers on the same plant.)

Pollination also requires consideration of pollenizers. The terms "pollinator" and "pollenizer" are often confused: a *pollinator* is the agent that moves the pollen, whether it be bees, flies, bats, moths, or birds; a *pollenizer* is the plant that serves as the pollen source for other plants. Some plants are *self-compatible* (*self-fertile*) and can pollinate and fertilize themselves. Other plants have chemical or physical barriers to self-pollination.

In agriculture and horticulture pollination management, a good pollenizer is a plant that provides compatible, viable and plentiful pollen and blooms at the same time as the plant that is to be pollinated or has pollen that can be stored and used when needed to pollinate the desired flowers. Hybridization is effective pollination between flowers of different species, or between different breeding lines or populations.

Peaches are considered self-fertile because a commercial crop can be produced without cross-pollination, though cross-pollination usually gives a better crop. Apples are considered self-incompatible, because a commercial crop must be cross-pollinated. Many commercial fruit tree varieties are grafted clones, genetically identical. An orchard block of apples of one variety is genetically a single plant. Many growers now consider this a mistake. One means of correcting this mistake is to graft a limb of an appropriate pollenizer (generally a variety of crabapple) every six trees or so.

The wasp *Mischocyttarus* rotundicollis transporting pollen grains of *Schinus terebinthifolius*

Pollen Vectors

Biotic pollen vectors are animals, usually insects, but also reptiles, birds, mammals, and sundry others, that routinely transport pollen and play a role in pollination. This is usually as a result of their activities when visiting plants for feeding, breeding or shelter. The pollen adheres to the vector's body parts such as face, legs, mouthparts, hair, feathers, and moist spots; depending on the particular vector. Such transport is vital to the pollination of many plant species.

Any kind of animal that often visits or encounters flowers is likely to be a pollen vector to some extent. For example, a crab spider that stops at one flower for a time and then moves on, might carry pollen incidentally, but most pollen vectors of significant interest are those that routinely visit the flowers for some functional activity. They might feed on pollen, or plant organs, or on plant secretions such as nectar, and carry out acts of pollination on the way. Many plants bear flowers that favour certain types of pollinator over all others. This need not always be an effective strategy, because some flowers that are of such a shape that they favor pollinators that pass by their anthers and stigmata on the way to the nectar, may get robbed by ants that are small enough to bypass the normal channels, or by short-tongued bees that bite through the bases of deep corolla tubes to extract nectar at the end opposite to the anthers and stigma. Some flowers have specialized mechanisms to trap pollinators to increase effectiveness. Other flowers will attract pollinators by odor. For example, bee species such as *Euglossa cordata* are attracted to orchids this way, and it has been suggested that the bees will become intoxicated during these visits to the orchid flowers, which last up to 90 minutes. However, in general, plants that rely on pollen vectors tend to be adapted to their particular type of vector, for example day-pollinated species tend to be brightly coloured, but if they are pollinated largely by birds or specialist mammals, they tend to be larger and have larger nectar rewards than species that are strictly insect-pollinated. They also tend to spread their rewards over longer periods, having long flowering seasons; their specialist pollinators would be likely to starve if the pollination season were too short.

As for the types of pollinators, reptile pollinators are known, but they form a minority in most ecological situations. They are most frequent and most ecologically significant in island systems, where insect and sometimes also bird populations may be unstable and less species-rich. Adaptation to a lack of animal food and of predation pressure, might therefore favour reptiles becoming more herbivorous and more inclined to feed on pollen and nectar. Most species of lizards in the families that seem to be significant in pollination seem to carry pollen only incidentally, especially the larger species such as Varanidae and Iguanidae, but especially several species of the Gekkonidae are active pollinators, and so is at least one species of the Lacertidae, *Podarcis lilfordi*, which pollinates various species, but in particular is the major pollinator of *Euphorbia dendroides* on various Mediterranean islands.

Mammals are not generally thought of as pollinators, but some rodents, bats and marsupials are significant pollinators and some even specialise in such activities. In South Africa certain species of *Protea* (in particular *Protea humiflora, P. amplexicaulis, P. subulifolia, P. decurrens* and *P. cordata*) are adapted to pollination by rodents (particularly Cape Spiny Mouse, *Acomys subspinosus*) and elephant shrews (*Elephantulus* species). The flowers are borne near the ground, are yeasty smelling, not colourful, and sunbirds reject the nectar with its high xylose content. The mice apparently can digest the xylose and they eat large quantities of the pollen. In Australia pollination by flying, gliding and earthbound mammals has been demonstrated.

Examples of pollen vectors include many species of wasps, that transport pollen of many plant species, being potential or even efficient pollinators.

Evolution of Plant/Pollinator Interactions

The first fossil record for abiotic pollination is from fern-like plants in the late Carboniferous period. Gymnosperms show evidence for biotic pollination as early as the Triassic period. Many fossilized pollen grains show characteristics similar to the biotically dispersed pollen today. Furthermore, the gut contents, wing structures, and mouthpart morphologies of fossilized beetles and flies suggest that they acted as early pollinators. The association between beetles and angiosperms during the early Cretaceous period led to parallel radiations of angiosperms and insects into the late Cretaceous. The evolution of nectaries in late Cretaceous flowers signals the beginning of the mutualism between hymenopterans and angiosperms.

Bees provide a good example of the mutualism that exists between hymenopterans and angiosperms. Flowers provide bees with nectar (an energy source) and pollen (a source of protein). When bees go from flower to flower collecting pollen they are also depositing pollen grains onto the flowers, thus pollinating them. While pollen and nectar, in most cases, are the most notable reward attained from flowers, bees also visit flowers for other resources such as oil, fragrance, resin and even waxes. It has been estimated that bees originated with the origin or diversification of angiosperms. In addition, cases of coevolution between bee species and flowering plants have been illustrated by specialized adaptations. For example, long legs are selected for in *Rediviva neliana*, a bee that collects oil from *Diascia capsularis*, which have long spur lengths that are selected for in order to deposit pollen on the oil-collecting bee, which in turn selects for even longer legs in *R. neliana* and again longer spur length in *D. capsularis* is selected for, thus, continually driving each other's evolution.

In Agriculture

An *Andrena* bee collects pollen among the stamens of a rose. The female carpel structure appears rough and globular to the left. The bee's stash of pollen is on its hind leg.

Bombus ignitus, a popular commercial pollinator in Japan and China

Blueberries being pollinated by bumblebees. Bumblebee hives need to be bought each year as the queens must hibernate (unlike honey bees). They are used nonetheless as they offer advantages with certain fruits as blueberries (such as the fact that they are active even at colder outdoor ambient temperature).

Well-pollinated blackberry blossom begins to develop fruit. Each incipient drupelet has its own stigma and good pollination requires the delivery of many grains of pollen to the flower so that all drupelets develop.

Pollination management is a branch of agriculture that seeks to protect and enhance present pollinators and often involves the culture and addition of pollinators in monoculture situations, such as commercial fruit orchards. The largest managed pollination event in the world is in Californian almond orchards, where nearly half (about one million hives) of the US honey bees are trucked to the almond orchards each spring. New York's apple crop requires about 30,000 hives; Maine's blueberry crop uses about 50,000 hives each year.

Bees are also brought to commercial plantings of cucumbers, squash, melons, strawberries, and many other crops. Honey bees are not the only managed pollinators: a few other species of bees are also raised as pollinators. The alfalfa leafcutter bee is an important pollinator for alfalfa seed in western United States and Canada. Bumblebees are increasingly raised and used extensively for greenhouse tomatoes and other crops.

The ecological and financial importance of natural pollination by insects to agricultural crops, improving their quality and quantity, becomes more and more appreciated and has given rise to new financial opportunities. The vicinity of a forest or wild grasslands with native pollinators near agricultural crops, such as apples, almonds or coffee can improve their yield by about 20%. The benefits of native pollinators may result in forest owners demanding payment for their contribution in the improved crop results – a simple example of the economic value of ecological services. Farmers can also raise native crops in order to promote native bee pollinator species as shown with *L. vierecki* in Delaware and *L. leucozonium* in southwest Virginia.

The American Institute of Biological Sciences reports that native insect pollination saves the United States agricultural economy nearly an estimated $3.1 billion annually through natural crop production; pollination produces some $40 billion worth of products annually in the United States alone.

Pollination of food crops has become an environmental issue, due to two trends. The trend to monoculture means that greater concentrations of pollinators are needed at bloom time than ever before, yet the area is forage poor or even deadly to bees for the rest of the season. The other trend is the decline of pollinator populations, due to pesticide misuse and overuse, new diseases and parasites of bees, clearcut logging, decline of beekeeping, suburban development, removal of hedges and other habitat from farms, and public concern about bees. Widespread aerial spraying for mosquitoes due to West Nile fears is causing an acceleration of the loss of pollinators.

The US solution to the pollinator shortage, so far, has been for commercial beekeepers to become pollination contractors and to migrate. Just as the combine harvesters follow the wheat harvest from Texas to Manitoba, beekeepers follow the bloom from south to north, to provide pollination for many different crops.

In some situations, farmers or horticulturists may aim to restrict natural pollination to only permit breeding with the preferred individuals plants. This may be achieved through the use of pollination bags.

Improving Pollination in Areas with Suboptimal Bee Densities

In some instances growers' demand for beehives far exceeds the available supply. The number of managed beehives in the US has steadily declined from close to 6 million after WWII, to less than 2.5 million today. In contrast, the area dedicated to growing bee-pollinated crops has grown over 300% in the same time period. Additionally, in the past five years there has been a decline in winter managed beehives, which has reached an unprecedented rate of colony losses at near 30%. At present, there is an enormous demand for beehive rentals that cannot always be met. There is a clear need across the agricultural industry for a management tool to draw pollinators into cultivations and encourage them to preferentially visit and pollinate the flowering crop. By attracting pollinators like honeybees and increasing their foraging behavior, particularly in the center of large plots, we can increase grower returns and optimize yield from their plantings. ISCA Technologies, from Riverside California, created a semiochemical formulation called SPLAT Bloom, that modifies the behavior of honeybees, inciting them to visit flowers in every portion of the field.

Environmental Impacts

Loss of pollinators, also known as Pollinator decline (of which colony collapse disorder is perhaps the most well known) has been noticed in recent years. Observed losses would have significant economic impacts. Possible explanations for pollinator decline include habitat destruction, pesticide, parasitism/diseases, climate change and others, and many researchers believe it is the synergistic effects of these factors which are ultimately detrimental to pollinator populations.

The Structure of Plant-pollinator Networks

Wild pollinators often visit a large number of plant species and plants are visited by a large number of pollinator species. All these relations together form a network of interactions between plants and pollinators. Surprising similarities were found in the structure of networks consisting out of the interactions between plants and pollinators. This structure was found to be similar in very different ecosystems on different continents, consisting of entirely different species.

The structure of plant-pollinator networks may have large consequences for the way in which pollinator communities respond to increasingly harsh conditions. Mathematical models, examining the consequences of this network structure for the stability of pollinator communities suggest that the specific way in which plant-pollinator networks are organized minimizes competition between pollinators and may even lead to strong indirect facilitation between pollinators when conditions are harsh. This means that pollinator species together can survive under harsh conditions. But it also means that pollinator species collapse simultaneously when conditions pass a critical point. This simultaneous collapse occurs, because pollinator species depend on each other when surviving under difficult conditions.

Such a community-wide collapse, involving many pollinator species, can occur suddenly when increasingly harsh conditions pass a critical point and recovery from such a collapse might not be easy. The improvement in conditions needed for pollinators to recover, could be substantially larger than the improvement needed to return to conditions at which the pollinator community collapsed.

Forms of Pollination

Anemophily

Wind-pollination (anemophily) syndrome

The flowers of wind-pollinated flowering plants, such as this saw-tooth oak (*Quercus acutissima*), are less showy than insect-pollinated flowers.

Anemophilous plants, such as this pine (*Pinus*) produce large quantities of pollen, which is carried on the wind.

Anemophily or wind pollination is a form of pollination whereby pollen is distributed by wind. Almost all gymnosperms are anemophilous, as are many plants in the order Poales, including grasses, sedges and rushes. Other common anemophilous plants are oaks, sweet chestnuts, alders and members of the family Juglandaceae (hickory or walnut family).

Syndrome

Features of the wind-pollination syndrome include a lack of scent production, a lack of showy floral parts (resulting in inconspicuous flowers), reduced production of nectar, and the production of enormous numbers of pollen grains. This distinguishes them from entomophilous and zoophilous species (whose pollen is spread by insects and vertebrates respectively).

Anemophilous pollen grains are light and non-sticky, so that they can be transported by air currents. They are typically 20–60 micrometres (0.0008–0.0024 in) in diameter, although the pollen grains of *Pinus* species can be much larger and much less dense. Anemophilous plants possess well-exposed stamens so that the pollens are exposed to wind currents and also have large and feathery stigma to easily trap airborne pollen grains. Pollen from anemophilous plants tends to be smaller and lighter than pollen from entomophilous ones, with very low nutritional value to insects. However, insects sometimes gather pollen from staminate anemophilous flowers at times when higher-protein pollens from entomophilous flowers are scarce. Anemophilous pollens may also be inadvertently captured by bees' electrostatic field. This may explain why, though bees are not observed to visit ragweed flowers, its pollen is often found in honey made during the ragweed floral bloom. Other flowers that are generally anemophilous are observed to be actively worked by bees, with solitary bees often visiting grass flowers, and the larger honeybees and bumblebees frequently gathering pollen from corn tassels and other grains.

Anemophily is an adaptation that helps to separate the male and female reproductive systems of a single plant, reducing the effects of inbreeding. It often accompanies dioecy – the presence of male and female reproductive structures on separate plants.

Allergies

Almost all pollens that are allergens are from anemophilous species. Grasses (Poaceae) are the most important producers of aeroallergens in most temperate regions, with lowland or meadow species producing more pollen than upland or moorland species.

Hydrophily

Hydrophily is a fairly uncommon form of pollination whereby pollen is distributed by the flow of waters, particularly in rivers and streams. Hydrophilous species fall into two categories: those that distribute their pollen to the surface of water, and those that distribute it beneath the surface.

Surface Pollination

Surface pollination is more frequent, and appears to be a transitional phase between wind pollination and true hydrophily. In these the pollen floats on the surface and reaches the stigmas of the female flowers as in *Hydrilla, Callitriche, Ruppia, Zostera, Elodea*. In *Vallisneria* the male flowers become detached and float on the surface of the water; the anthers are thus brought in contact with the stigmas of the female flowers. Surface hydrophily has been observed in several species of *Potamogeton* as well as some marine species.

Submerged Pollination

Species exhibiting true submerged hydrophily include *Najas*, where the pollen grains are heavier than water, and sinking down are caught by the stigmas of the extremely simple female flowers, *Posidonia australis* and *Zostera marina*.

Entomophily

Bee pollinating a flower

Entomophily or insect pollination is a form of pollination whereby pollen of plants, especially but not only of flowering plants, is distributed by insects. Flowers pollinated by insects typically advertise themselves with bright colours, sometimes with conspicuous patterns (honey guides) leading to rewards of pollen and nectar; they may also have an attractive scent which in some cases mimics insect pheromones. Insect pollinators such as bees have adaptations for their role, such as lapping or sucking mouthparts to take in nectar, and in some species also pollen baskets on their hind legs. This required the coevolution of insects and flowering plants in the development of pollination behaviour by the insects and pollination mechanisms by the flowers, benefiting both groups.

Soldier beetle covered with pollen

Many plants, including flowering plants such as grasses, are instead pollinated by other mechanisms, such as by wind.

Coevolution

History

The insect-pollinated flowers of angiosperms use a combination of cues such as bright colours and streaked patterns to advertise themselves to insects.

The early spermatophytes (seed plants) were largely dependent on the wind to carry their pollen from one plant to another, and it was around 125 to 115 million years ago that a new pollination strategy developed and angiosperms (flowering plants) first appeared. Before that, insect involvement in pollination was limited to "pollination assistants", insects which inadvertently carried the pollen between plants merely by their movements. The real relationship between plants and insects began with the arrival of the first angiosperms in the Early Cretaceous. The morphology of the first fossil basal angiosperms has similarity to modern-day plants that are fertilised by beetles.

It seems likely that beetles led the way in insect pollination, followed by flies. Among the twelve living families of basal angiosperms, six are predominantly pollinated by flies, five by beetles and only one by bees. Nevertheless, traits such as sapromyophily (emitting the odour of carrion to attract flies) have evolved independently in several unrelated angiosperm families.

The Plant's Needs

Wind and water pollination require the production of vast quantities of pollen because of the chancy nature of its deposition. If they are not to be reliant on the wind or water (for aquatic species), plants need pollinators to move their pollen grains from one plant to another. They particularly need pollinators to consistently choose flowers of the same species, so they have evolved different lures to encourage specific pollinators to maintain fidelity to the same species. The attractions offered are mainly nectar, pollen, fragrances and oils. The ideal pollinating insect is hairy (so that pollen adheres to it), and spends time exploring the flower so that it comes into contact with the reproductive structures.

Mechanisms

Many insects are pollinators, particularly bees, Lepidoptera (butterflies and moths), wasps, flies, ants and beetles. On the other hand, some plants are generalists, being pollinated by insects in several orders. Entomophilous plant species have frequently evolved mechanisms to make themselves more appealing to insects, e.g., brightly coloured or scented flowers, nectar, or appealing shapes and patterns. Pollen grains of entomophilous plants are generally larger than the fine pollens of anemophilous (wind-pollinated) plants, which has to be produced in much larger quantities because such a high proportion is wasted. This is energetically costly, but in contrast, entomophilous plants have to bear the energetic costs of producing nectar.

Hummingbird moth on *Clarkia*

Butterflies and moths have hairy bodies and long proboscides which can probe deep into tubular flowers. Butterflies mostly fly by day and are particularly attracted to pink, mauve and purple flowers. The flowers are often large and scented, and the stamens are so-positioned that pollen is deposited on the insects while they feed on the nectar. Moths are mostly nocturnal and are attracted by night-blooming plants. The flowers of these are often tubular, pale in colour and fragrant only at night. Hawkmoths tend to visit larger flowers and hover as they feed; they transfer pollen by means of the proboscis. Other moths land on the usually smaller flowers, which may be aggregated into flowerheads. Their energetic needs are not so great as those of hawkmoths and they are offered smaller quantities of nectar.

Inflorescences pollinated by beetles tend to be flat with open corollas or small flowers clustered in a head with multiple, projecting anthers that shed pollen readily. The flowers are often green or pale-coloured, and heavily scented, often with fruity or spicy aromas, but sometimes with odours of decaying organic matter. Some, like the giant water lily, include traps designed to retain the beetles in contact with the reproductive parts for longer periods.

Female hoverfly *Dasysyrphus albostriatus*

Unspecialised flies with short proboscides are found visiting primitive flowers with readily accessible nectar. More specialised flies like syrphids and tabanids can visit more advanced blooms, but their purpose is to nourish themselves, and any transfer of pollen from one flower to another happens haphazardly. The small size of many flies is often made up for by their abundance, however they are unreliable pollinators as they may bear incompatible pollen, and lack of suitable breeding habitats may limit their activities. Some *Pterostylis* orchids are pollinated by midges unique to each species. A decline, for whatever reason, to one side of this partnership can be catastrophic for the other.

Flowers pollinated by bees and wasps vary in shape, colour and size. Yellow or blue plants are often visited, and flowers may have ultra-violet nectar guides, that help the insect to find the nectary. Some flowers, like sage or pea, have lower lips that will only open when sufficiently heavy insects, such as bees, land on them. With the lip depressed, the anthers may bow down to deposit pollen on the insect's back. Other flowers, like tomato, may only liberate their pollen by buzz pollination, a technique in which a bumblebee will cling on to a flower while vibrating its flight muscles, and this dislodges the pollen. Because bees care for their brood, they need to collect more food than just to maintain themselves, and therefore are important pollinators. Other bees are nectar thieves and bite their way through the corolla in order to raid the nectary, in the process bypassing the reproductive structures.

Ants are not well adapted to pollination but they have been shown to perform this function in *Polygonum cascadense* and in certain desert plants with small blossoms near the ground with little fragrance or visual attraction, small quantities of nectar and limited quantities of sticky pollen.

Plant-Insect Pairings

The bee orchid mimics bees in appearance and scent, implying close coevolution of a species of flower and a species of insect.

Some plant species co-evolved with a particular pollinator species, such as the bee orchid. The species is almost exclusively self-pollinating in its northern ranges, but is pollinated by the solitary bee *Eucera* in the Mediterranean area. The plant attracts these insects by producing a scent that mimics the scent of the female bee. In addition, the lip acts as a decoy, as the male bee confuses it with a female that is visiting a pink flower. Pollen transfer occurs during the ensuing pseudocopulation.

Cross section of a *Ficus glomerata* (fig) fruit showing the syconium with pollinating fig wasps inside.

Figs in the genus *Ficus* have a mutualistic arrangement with certain tiny agaonid wasps. In the common fig, the inflorescence is a syconium, formed by an enlarged, fleshy, hollow receptacle with multiple ovaries on the inner surface. A female wasp enters through a narrow aperture, fertilises these pistillate flowers and lays its eggs in some ovaries, with galls being formed by the developing larvae. In due course, staminate flowers develop inside the syconium. Wingless male wasps hatch and mate with females in the galls before tunnelling their way out of the developing fruit. The winged females, now laden with pollen, follow, flying off to find other receptive syconia at the right stage of development. Most species of fig have their own unique commensal species of wasp.

Taxonomic Range

Wind pollination is the reproductive strategy adopted by the grasses, sedges, rushes and catkin-bearing plants. Other flowering plants are mostly pollinated by insects (or birds or bats), which seems to be the primitive state, and some plants have secondarily developed wind pollination. Some plants that are wind pollinated have vestigial nectaries, and other plants like common heather that are regularly pollinated by insects, produce clouds of pollen and some wind pollination is inevitable. The hoary plantain is primarily wind pollinated, but is also visited by insects which pollinate it. In general, showy, colourful, fragrant flowers like sunflowers, orchids and *Buddleja* are insect pollinated. The only entomophilous plants that are not seed plants are the dung-mosses of the family Splachnaceae.

Zoophily

A Rufous Hummingbird (*Selasphorus rufus*) is attracted to brightly colored flowers and assists the pollination of the plant.

Zoophily is a form of pollination whereby pollen is transferred by vertebrates, particularly by hummingbirds and other birds, and bats, but also by monkeys, marsupials, lemurs, bears, rabbits, deer, rodents, lizards and other animals. Zoomophilous species, like entomophilous species, frequently evolve mechanisms to make themselves more appealing to the particular type of pollinator, e.g. brightly colored or scented flowers, nectar, and appealing shapes and patterns. These plant animal relationships are often mutually beneficial because of the food source provided in exchange for pollination. Zoophilous species include *Arctium*, *Acaena*, and *Galium aparine'* Pollination is defined as the transfer of pollen from the anther to the stigma (Worldnet). There are many vectors for pollination, including abiotic (wind and water), and biotic (animal). There are different benefits and costs associated with any vector type. For instance, using animal pollination is beneficial because the process is more directed and often results in pollination. At the same time it is costly for the plant to produce rewards, such as nectar, to attract animal pollinators. Not producing such rewards is one benefit of using abiotic pollinators, but a cost associated with this approach is that the pollen may be distributed somewhat randomly. In general, pollination by vertebrates occurs when the animal reaches inside the flowers for nectar. While feeding on the nectar, the animal rubs or touches the stamens and is covered in pollen. Some of this pollen will be deposited on the stigma of the next flower it visits, pollinating the flower (Missouri Botanical Garden 2006).

Bat Pollination

Bat pollination is chiropterophily. Most bat species that pollinate flowers inhabit Africa, Southeast

Asia, and the Pacific Islands, although bat pollination occurs over a geographically wide range. Many flowers are dependent on bats for pollination, such as the flowers which produce mangoes, bananas, and guavas (Celebrating Wildlife 2006). Bat pollination is an integral process in tropical communities with 500 tropical plant species completely, or partially, dependent on bats for pollination (Heithaus 1974). Also, it has been noted that outcrossing (introducing unrelated genetic material into a breeding line) by bats increases genetic diversity and is important in tropical communities (Heithaus 1974).

Plants pollinated by bats often have white or pale nocturnal flowers that are large and bell shaped. Many of these flowers have large amounts of nectar, and emit a smell that attracts bats, such as a strong fruity or musky odor (Gibson 2001). Bats use certain chemical cues to locate food sources. They are attracted to odors that contain esters, alcohols, aldehydes, and aliphatic acids (Gibson 2001).

The banana bat (*Musonycteris harrisoni*) is a nectarivorous species found only on the Pacific coast of Mexico. It has a very small geographic range and is distinguishable by its extremely long nose. The long snout and tongue, one tongue recorded as measuring 76mm, allows this bat to feed on the nectar of long tubular flowers. This bat species is small, with the head and body length ranging from 70 to 79mm. The wild banana flower is elongated with a purple color (Tellez 1999).

Pollination by Other Mammals

Non-flying mammals (to distinguish them from bats) have been found to feed on the nectar of several species of plant. Though some of these mammals are pollinators, others do not carry or transfer enough pollen to be considered pollinators (Johnson 2001). This group of non-flying pollinators is mainly composed of marsupials, primates, and rodents (Johnson 2001). Well-documented studies of non-flying mammal pollination now involve at least 59 species of mammal distributed among 19 families and six orders (Carthewa 1997). As of 1997, there were 85 species of plants from 43 genera and 19 families which were visited by these mammals (Carthewa 1997). In many cases, a plant species is visited by a range of mammals. Two examples of multiple mammal pollination are the genus *Quararibea* which is visited by 12 species and *Combretum* which is visited by 8 (Carthewa 1997).

Plant species that feed non-flying mammals will often exhibit similar characteristics to aid in pollination. The flowers are often large and sturdy, or are grouped together as multi-flowered inflorescences. Many non-flying mammals are nocturnal and have an acute sense of smell, so the plants tend not to have bright showy colors, but instead excrete a pungent odor. Plants will often flower profusely and produce a large amount of sugar-rich nectar. These plants also tend to produce large amounts of pollen because mammals are larger than some other pollinators, and lack the precision smaller pollinators can achieve (Carthewa 1997). Animals with more precision, such as bees or other insects with a proboscis, can pollinate small flowers with less pollen necessary. This means that a plant will require more pollen for a larger mammal pollinator.

One example of a symbiotic relationship between a plant and its animal pollinators is the African Lily, *Massonia depressa*, and some rodent species of the Succulent Karoo region of South Africa. At least four rodent species, including two gerbil species, were found to be visiting *M. depressa* during the night (Johnson 2001). Traits of the *M. depressa* flowers support non-flying mammal pollination. It has dull-colored and very sturdy flowers at ground level, has a strong yeasty odor, and secretes copious amounts of sucrose-dominant nectar during the night (Johnson 2001). The nectar of *M. depressa* was also found to be 400 times as viscous, or resistant to flow, as an equivalent sugar solution. This jelly-like consistency of the nectar may discourage insect consumption while also facilitating lapping by rodents. It is assumed that *M. depressa* coevolved with its pollinators (Johnson 2001).

Bird Pollination

The term ornithophily is used to describe pollination specifically by birds. Hummingbirds, found only in North and South America, are the most recognized nectar-eating bird, but there are many other bird species throughout the world that are also important pollinators. These include: sunbirds, honeyeaters, flowerpeckers, honeycreepers, and bananaquits (University of Connecticut 2006).

Plants pollinated by birds often have brightly colored diurnal flowers that are red, yellow, or orange, but no odor because birds have a poor sense of smell. Other characteristics of these plants are that they have suitable, sturdy places for perching, abundant nectar that is deeply nested within the flower. Often flowers are elongated or tube shaped. Also, many plants have anthers placed in the flower so that pollen rubs against the birds head/back as the bird reaches in for nectar (Celebrating Wildlife 2006).

The ruby-throated hummingbird (*Archilochus colubris*) is one of many species of hummingbirds. Found in North and Central America, this bird is an important pollinator for a variety of plant species. Some species, such as the trumpet creeper, are adapted specifically for ruby-throated hummingbirds (Harris 2000). This species is quite small, measuring 7.5-9.0 cm long and weighing only 3.4-3.8g. The long narrow bill of the hummingbird is the perfect tool for extracting nectar from elongated flowers (Harris 2000). This species is attracted to brightly colored flowers, especially those that are red in color (Harris 2000).

Lizard Pollination

Although lizard pollination has historically been underestimated, recent studies have shown lizard pollination to be an important part of many plant species' survival. Not only do lizards show mutualistic relationships, but these are found to occur most often on islands. This pattern of lizard pollination on islands is mainly due to their high densities, a surplus of floral food, and a relatively low predation risk when compared to lizards on the mainland (Olesen 2003).

The lizard *Hoplodactylus* is only attracted by nectar on flowers, not pollen. This means flowers pollinated by this species must produce copious nectar as a reward for *Hoplodactylus*. Scented flowers are another important adaptation to attract lizards due to their acute sense of smell. Although lizards have the ability to distinguish colors, as nocturnal feeders, it is more difficult to see bright colors. Because *Hoplodactylus* feeds nocturnally, it is sometimes less important for flowers to allocate resources to showy inflorescences. Flowers must also be robust enough to support the weight of the pollinator while feeding (Whitaker 1987).

In New Zealand, *Hoplodactylus* geckos visit flowers of many native plant species for nectar and pollen. The flowers of *Metrosideros excelsa* are pollinated by more than 50 types of gecko as well as birds and bees. Of the geckos visiting this species, two-thirds of them carried large amounts of pollen, suggesting a main role in pollination. However, after the arrival of humans in New Zealand, lizard populations have declined making it more difficult to witness lizard pollination (Olesen 2003).

Open Pollination

Detasseling corn (maize) plants from one variety in a field where two varieties are planted. The male flowers are removed so that all seeds are hybrids sired from the second variety.

The terms "open pollination" and "open pollinated" refer to a variety of concepts in the context of the sexual reproduction of plants.

True-breeding Definition

"Open pollinated" generally refers to seeds that will "breed true." When the plants of an open-pollinated variety self-pollinate, or are pollinated by another representative of the same variety, the resulting seeds will produce plants roughly identical to their parents. This is in contrast to the seeds produced by plants that are the result of a recent cross (such as, but not confined to, an F1 hybrid), which are likely to show a wide variety of differing characteristics. Open-pollinated varieties are also often referred to as standard varieties or, when the seeds have been saved across generations or across several decades, heirloom varieties. While heirlooms are usually open-pollinated, open-pollinated seeds are not necessarily heirlooms; open-pollinated varieties are still being developed.

One of the challenges in maintaining an open-pollinated variety is avoiding introduction of pollen from other strains. Based on how broadly the pollen for the plant tends to disperse, it can be controlled to varying degrees by greenhouses, tall wall enclosures, field isolation, or other techniques.

Because they breed true, the seeds of open-pollinated plants are often saved by home gardeners and farmers. Popular examples of open-pollinated plants include heirloom tomatoes, beans, peas, and many other garden vegetables.

Uncontrolled Pollination Definition

A second use of the term "open pollination" refers to pollination by insects, birds, wind, or other natural mechanisms. This can be contrasted with cleistogamy, closed pollination, which is one of the many types of self pollination. When used in this sense, open pollination may contrast with controlled pollination, a procedure used to ensure that all seeds of a crop are descended from parents with known traits, and are therefore more likely to have the desired traits.

The seeds of open-pollinated plants will produce new generations of those plants; however, because breeding is uncontrolled and the pollen (male parent) source is unknown, open pollination may result in plants that vary widely in genetic traits. Open pollination may increase biodiversity.

Some plants (such as many crops) are primarily self pollenizing and also breed true, so that even under open pollination conditions the next generation will be (almost) the same. Even among true breeding organisms, some variation due to genetic recombination or to mutation can produce a few "off types".

Relationship to Hybridization

Hybrid pollination, a type of controlled pollination in which the pollen comes from a different strain (or species), can be used to increase crop suitability, especially through heterosis. The resulting hybrid strain can sometimes be inbred and selected for desired traits until a strain that breeds true by open pollination is achieved. The result is referred to as a inbred hybrid strain. To add some confusion, the term hybrid inbred applies to hybrids that are made from selected inbred lines that have certain desired characteristics. The latter type of hybrid is sometimes designated F1 hybrid, i.e. the first hybrid (filial) generation whose parents were (different) inbred lines.

Pollenizer

A pollenizer or polleniser, sometimes pollinizer or polliniser is a plant that provides pollen.

The word pollinator is often used when pollenizer is more precise. A pollinator is the biotic agent that moves the pollen, such as bees, moths, bats, and birds. Bees are thus often referred to as 'pollinating insects'.

The verb form to pollenize is to be the source of pollen, or to be the sire of the next plant generation.

While some plants are capable of self pollenization, the term is more often used in pollination management for a plant that provides abundant, compatible, and viable pollen at the same flowering time as the pollinated plant. For example, most crabapple varieties are good pollenizers for any apple tree that blooms at the same time, and are often used in apple orchards for the purpose. Some apple cultivars produce very little pollen or pollen that is sterile or incompatible with other apple varieties. These are poor pollenizers.

A pollenizer can also be the male plant in dioecious species (where entire plants are of a single sex), such as with kiwifruit or holly.

Nursery catalogs often specify that cultivar X should be planted as a pollinator for cultivar Y, when they actually should be referring to it as a pollenizer. Strictly, a plant can only be a pollinator when it is self-fertile and it physically pollinates itself without the aid of an external pollinator.

Self-pollination

Self-pollination is when pollen from the same plant arrives at the stigma of a flower (in flowering plants) or at the ovule (in Gymnosperms). There are two types of self-pollination: In autogamy, pollen is transferred to the stigma of the same flower. In geitonogamy, pollen is transferred from the anther of one flower to the stigma of another flower on the same flowering plant, or from microsporangium to ovule within a single (monoecious) Gymnosperm. Some plants have mechanisms that ensure autogamy, such as flowers that do not open (cleistogamy), or stamens that move to come into contact with the stigma.

One type of automatic self-pollination occurs in the orchid *Ophrys apifera*. One of the two pollinia bends itself towards the stigma.

Occurrence

Few plants self-pollinate without the aid of pollen vectors (such as wind or insects). The mechanism is seen most often in some legumes such as peanuts. In another legume, soybeans, the flowers open and remain receptive to insect cross pollination during the day. If this is not accomplished, the flowers self-pollinate as they are closing. Among other plants that can self-pollinate are many kinds of orchids, peas, sunflowers and tridax. Most of the self-pollinating plants have small, relatively inconspicuous flowers that shed pollen directly onto the stigma, sometimes even before the bud opens. Self-pollinated plants expend less energy in the production of pollinator attractants and can grow in areas where the kinds of insects or other animals that might visit them are absent or very scarce—as in the Arctic or at high elevations.

Self-pollination limits the variety of progeny and may depress plant vigor. However, self-pollination can be advantageous, allowing plants to spread beyond the range of suitable pollinators or produce offspring in areas where pollinator populations have been greatly reduced or are naturally variable.

Pollination can also be accomplished by cross-pollination. Cross-pollination is the transfer of pollen, by wind or animals such as insects and birds, from the anther to the stigma of flowers on separate plants.

Types of Flowers that Self Pollinate

Both hermaphrodite and monoecious species have the potential for self-pollination leading to self-fertilization unless there is a mechanism to avoid it. Eighty percent of all flowering plants are hermaphroditic, meaning they contain both sexes in the same flower, while 5 percent of plant species are monoecious. The remaining 15% would therefore be dioecious (each plant unisexual).

Advantages of Self-pollination

There are several advantages for self-pollinating flowers. If a given genotype is well-suited for an environment, self-pollination helps to keep this trait stable in the species. Not being dependent on pollinating agents allows self-pollination to occur when bees and wind are nowhere to be found. Self-pollination can be an advantage when the number of flowers are small or widely spaced.

Disadvantages of Self-pollination

The disadvantages of self-pollination come from a lack of variation that allows no adaptation to the changing environment or potential pathogen attack. Self-pollination can lead to inbreeding depression, or the reduced health of the species, due to the breeding of related specimens. This is why many flowers that could potentially self-pollinate have a built-in mechanism to avoid it, or make it second choice at best. Genetic defects in self-pollinating plants cannot be eliminated by genetic recombination and offspring can only avoid inheriting the deleterious attributes through a chance mutation arising in a gamete.

Mixed Mating

About 42% of flowering plants exhibit a mixed mating system in nature. In the most common kind of system, individual plants produce a single flower type and fruits may contain self-pollinated, out-crossed or a mixture of progeny types. Another mixed mating system is referred to as dimorphic cleistogamy. In this system a single plant produces both open, potentially out-crossed and closed, obligately self-pollinated cleistogamous flowers.

Self-pollinating Species

The evolutionary shift from outcrossing to self-fertilization is one of the most common evolutionary transitions in plants. About 10-15% of flowering plants are predominantly self-fertilizing. A few well-studied examples of self-pollinating species are described below.

Paphiopedilum Parishii

Self-pollination in the slipper orchid *Paphiopedilum parishii* occurs when the anther changes from a solid to a liquid state and directly contacts the stigma surface without the aid of any pollinating agent.

Holcoglossum Amesianum

The tree-living orchid *Holcoglossum amesianum* has a type of self-pollination mechanism in which the bisexual flower turns its anther against gravity through 360° in order to insert pollen into its own stigma cavity---without the aid of any pollinating agent or medium. This type of self-pollination appears to be an adaptation to the windless, drought conditions that are present when flowering occurs, at a time when insects are scarce. Without pollinators for outcrossing, the necessity of ensuring reproductive success appears to outweigh potential adverse effects of inbreeding. Such an adaptation may be widespread among species in similar environments.

Caulokaempferia Coenobialis

In the Chinese herb *Caulokaempferia coenobialis* a film of pollen is transported from the anther (pollen sacs) by an oily emulsion that slides sideways along the flower's style and into the individual's own stigma. The lateral flow of the film of pollen along the style appears to be due solely to the spreading properties of the oily emulsion and not to gravity. This strategy may have evolved to cope with a scarcity of pollinators in the extremely shady and humid habitats of *C. coenobialis*.

Capsella Rubella

Capsella rubella (Red Shepard's purse) is a self-pollinating species that became self-compatible 50,000 to 100,000 years ago, indicating that self-pollination is an evolutionary adaptation that can persist over many generations. Its out-crossing progenitor was identified as *Capsella grandiflora*.

Arabidopsis Thaliana

Arabidopsis thaliana is a predominantly self-pollinating plant with an out-crossing rate in the wild estimated at less than 0.3%. A study suggested that self-pollination evolved roughly a million years ago or more.

Possible Long-term Benefit of Meiosis

Meiosis followed by self-pollination produces little overall genetic variation. This raises the question of how meiosis in self-pollinating plants is adaptively maintained over extended periods (i. e. for at least 10^4 to 10^6 years) in preference to a less complicated and less costly asexual ameiotic process for producing progeny. An adaptive benefit of meiosis that may explain its long-term maintenance in self-pollinating plants is efficient recombinational repair of DNA damage. This benefit can be realized at each generation (even when genetic variation is not produced).

Cleistogamy

Cleistogamy is a type of automatic self-pollination of certain plants that can propagate by using non-opening, self-pollinating flowers. Especially well known in peanuts, peas, and beans, this behavior is most widespread in the grass family. However, the largest genus of cleistogamous plants is actually *Viola*.

The more common opposite of cleistogamy, or "closed marriage," is called chasmogamy, or "open marriage." Virtually all plants that produce cleistogamous flowers also produce chasmogamous ones. The principal advantage of cleistogamy is that it requires less plant resources to produce seeds than does chasmogamy because development of petals, nectar and large amounts of pollen are not required. This efficiency makes cleistogamy particularly useful for seed production on unfavorable sites or adverse conditions. *Impatiens capensis,* for example, has been observed to produce only cleistogamous flowers after being severely damaged by grazing and to maintain populations on unfavorable sites with only cleistogamous flowers. The obvious disadvantage of cleistogamy is that self-fertilization occurs, which may suppress the creation of genetically superior plants.

For genetically modified (GM) rapeseed, researchers hoping to minimise the admixture of GM and non-GM crops are attempting to use cleistogamy to prevent gene flow. However, preliminary results from Co-Extra, a current project within the EU research program, show that although cleistogamy reduces gene flow, it is not at the moment a consistently reliable tool for biocontainment; due to a certain instability of the cleistogamous trait, some flowers may open and release genetically modified pollen.

Pollinator

A syrphid fly (*Eristalinus taeniops*) pollinating a common hawkweed

A pollinator is the biotic agent (vector) that moves pollen from the male anthers of a flower to the female stigma of a flower to accomplish fertilization or 'syngamy' of the female gametes in the ovule of the flower by the male gametes from the pollen grain. A pollinator is different from a pollenizer, which is a plant that is a source of pollen for the pollination process. *Anthecology* is the scientific study of pollination.

Insect pollinators include bees, (honey bees, solitary species, bumblebees); pollen wasps (Masarinae); ants; a variety of flies including bee flies and hoverflies; lepidopterans, both butterflies and moths; and flower beetles. Vertebrates, mainly bats and birds, but also some non-bat mammals (monkeys, lemurs, possums, rodents) and some reptiles (lizards and snakes) pollinate certain plants. Among the pollinating birds are hummingbirds, honeyeaters and sunbirds with long beaks; they pollinate a number of deep-throated flowers.

Cycads, which are not flowering plants, are also pollinated by insects.

Background

Plants fall into pollination syndromes that reflect the type of pollinator being attracted. These are characteristics such as: overall flower size, the depth and width of the corolla, the color (including patterns called nectar guides that are visible only in ultraviolet light), the scent, amount of nectar, composition of nectar, etc. For example, birds visit red flowers with long, narrow tubes and lots of nectar, but are not as strongly attracted to wide flowers with little nectar and copious pollen, which are more attractive to beetles. When these characteristics are experimentally modified (altering colour, size, orientation), pollinator visitation may decline.

It has recently been discovered that cycads, which are not flowering plants, are also pollinated by insects.

Types of Pollinators

Bees

Lipotriches sp. bee pollinating flowers

Honey bee with pollen adhering: Bees are the most effective insect pollinators.

The most recognized pollinators are the various species of bees, which are plainly adapted to pollination. Bees typically are fuzzy and carry an electrostatic charge. Both features help pollen grains adhere to their bodies, but they also have specialized pollen-carrying structures; in most bees, this takes the form of a structure known as the scopa, which is on the hind legs of most bees, and/or the lower abdomen (e.g., of megachilid bees), made up of thick, plumose setae. Honey bees, bumblebees, and their relatives do not have a scopa, but the hind leg is modified into a structure called the corbicula (also known as the "pollen basket"). Most bees gather nectar, a concentrated energy source, and pollen, which is high protein food, to nurture their young, and inadvertently transfer some among the flowers as they are working. Euglossine bees pollinate orchids, but these are male bees collecting floral scents rather than females gathering nectar or pollen. Female orchid bees act as pollinators, but of flowers other than orchids. Eusocial bees such as honey bees need an abundant and steady pollen source to multiply.

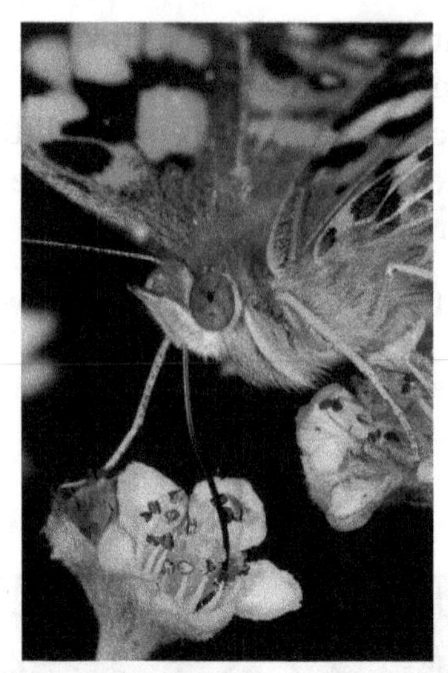

An Australian painted lady with its proboscis extended during feeding. Butterflies are recognised pollinators though not as effective as bees.

Scoliid wasp foraging

Tabanid on a thistle flower

Honey Bees

Honey bees travel from flower to flower, collecting nectar (later converted to honey), and pollen grains. The bee collects the pollen by rubbing against the anthers. The pollen collects on the hind legs, in a structure referred to as a "pollen basket". As the bee flies from flower to flower, some of the pollen grains are transferred onto the stigma of other flowers.

Nectar provides the energy for bee nutrition; pollen provides the protein. When bees are rearing large quantities of brood (beekeepers say hives are "building"), bees deliberately gather pollen to meet the nutritional needs of the brood.

Good pollination management seeks to have bees in a "building" state during the bloom period of the crop, thus requiring them to gather pollen, and making them more efficient pollinators. Thus, the management techniques of a beekeeper providing pollination services are different from, and to some extent in tension with, those of a beekeeper who is trying to produce honey.

Millions of hives of honey bees are contracted out as pollinators by beekeepers, and honey bees are by far the most important commercial pollinating agents, but many other kinds of pollinators, from blue bottle flies, to bumblebees, orchard mason bees, and leaf cutter bees are cultured and sold for managed pollination.

Other species of bees differ in various details of their behavior and pollen-gathering habits, and honey bees are not native to the Western Hemisphere; all pollination of native plants in the Americas historically has been performed by various native bees.

Other Insects

Many insects other than bees accomplish pollination by visiting flowers for nectar or pollen, or commonly both. Many do so adventitiously, but the most important pollinators are specialists for at least parts of their lifecycles for at least certain functions. For example, males of many species of Hymenoptera, including many hunting wasps, rely on freely flowering plants as sources of energy (in the form of nectar) and also as territories for meeting fertile females that visit the flowers. Prominent examples are predatory wasps (especially Sphecidae, Vespidae, and Pompilidae). The term "pollen wasps", in particular, is widely applied to the Masarinae, a subfamily of the Vespidae; they are remarkable among solitary wasps in that they specialise in gathering pollen for feeding their larvae, carried internally and regurgitated into a mud chamber prior to oviposition.

Many bee flies, and some Tabanidae and Nemestrinidae are particularly adapted to pollinating fynbos and Karoo plants with narrow, deep corolla tubes, such as *Lapeirousia* species. Part of the adaptation takes the form of remarkably long probosces.

Lepidoptera (butterflies and moths) also pollinate plants to various degrees. They are not major pollinators of food crops, but various moths are important pollinators of other commercial crops such as tobacco. Pollination by certain moths may be important, however, or even crucial, for some wildflowers mutually adapted to specialist pollinators. Spectacular examples include orchids such as *Angraecum sesquipedale*, dependant on a particular hawk moth, Morgan's sphinx. Yucca species provide other examples, being fertilised in elaborate ecological interactions with particular species of yucca moths.

Beetles of species that specialise in eating pollen, nectar, or flowers themselves, are important cross-pollinators of some plants such as members of the Araceae and Zamiaceae, that produce prodigious amounts of pollen. Others, for example the Hopliini, specialise in free-flowering species of the Asteraceae and Aizoaceae.

Various midges and thrips are comparatively minor opportunist pollinators. Ants also pollinate some kinds of flowers, but for the most part they are parasites, robbing nectar without conveying useful amounts of pollen to a stigma. Whole groups of plants, such as certain fynbos *Moraea* and *Erica* species produce flowers on sticky peduncles or with sticky corolla tubes that only permit access to flying pollinators, whether bird, bat, or insect.

Carrion flies and flesh flies in families such as Calliphoridae and Sarcophagidae are important for some species of plants whose flowers exude a fetid odor. The plants' ecological strategy varies; several species of *Stapelia*, for example, attract carrion flies that futilely lay their eggs on the flower, where their larvae promptly starve for lack of carrion. Other species do decay rapidly after ripening, and offer the visiting insects large masses of food, as well as pollen and sometimes seed to carry off when they leave.

Hoverflies are important pollinators of flowering plants worldwide. Often hoverflies are considered to be the second most important pollinators after wild bees. Although hoverflies as a whole are generally considered to be nonselective pollinators, some species have more specialized relationships. The orchid species *Epipactis veratrifolia* mimics alarm pheromones of aphids to attract hover flies for pollination. Another plant, the slipper orchid in southwest China, also achieves pollination by deceit by exploiting the innate yellow colour preference of syrphide.

Some male *Bactrocera* fruit flies are exclusive pollinators of some wild *Bulbophyllum* orchids that lack nectar and have a specific chemical attractant and reward (methyl eugenol, raspberry ketone or zingerone) present in their floral fragrances.

A class of strategy of great biological interest is that of sexual deception, where plants, generally orchids, produce remarkably complex combinations of pheromonal attractants and physical mimicry that induce male bees or wasps to attempt to mate with them, conveying pollinia in the process. Examples are known from all continents apart from Antarctica, though Australia appears to be exceptionally rich in examples.

Some Diptera (flies) may be the main pollinators at higher elevations of mountains, whereas *Bombus* species are the only pollinators among Apoidea in alpine regions at timberline and beyond.

Other insect orders are rarely pollinators, and then typically only incidentally (e.g., Hemiptera such as Anthocoridae and Miridae).

Vertebrates

Bats are important pollinators of some tropical flowers. Birds, particularly hummingbirds, honeyeaters and sunbirds also accomplish much pollination, especially of deep-throated flowers. Other vertebrates, such as kinkajous, monkeys, lemurs, possums, rodents and lizards have been recorded pollinating some plants.

Humans can be pollinators, as many gardeners have discovered that they must hand pollinate garden vegetables, whether because of pollinator decline (as has been occurring in parts of the U.S. since the mid-20th century) or simply to keep a strain genetically pure. This can involve using a small brush or cotton swab to move pollen, or to simply tap or shake tomato blossoms to release the pollen for the self pollinating flowers. Tomato blossoms are self-fertile, but (with the exception of potato-leaf varieties) have the pollen inside the anther, and the flower requires shaking to release the pollen through pores. This can be done by wind, by humans, or by a sonicating bee (one that vibrates its wing muscles while perched on the flower), such as a bumblebee. Sonicating bees are extremely efficient pollinators of tomatoes, and colonies of bumblebees are quickly replacing humans as the primary pollinators for greenhouse tomatoes.

Pollinator Population Declines and Conservation

Pollinators provide a key ecosystem service vital to the maintenance of both wild and agricultural plant communities. In 1999 the Convention on Biological Diversity issued the São Paulo Declaration on Pollinators, recognizing the critical role that these species play in supporting and maintaining terrestrial productivity as well as the survival challenges they face due to anthropogenic change. Today pollinators are considered to be in a state of decline; some species, such as Franklin's bumble bee (*Bombus franklini*) have been red-listed and are in danger of extinction. Although managed bee hives are increasing worldwide, these can not compensate for the loss of wild pollinators in many locations.

Declines in the health and population of pollinators pose what could be a significant threat to the integrity of biodiversity, to global food webs, and to human health. At least 80% of our world's crop species require pollination to set seed. An estimated one out of every three bites of food comes to us through the work of animal pollinators. The quality of pollinator service has declined over time and this had led to concerns that pollination will be less resistant to extinction in the future.

United States National Strategy to Promote the Health of Honey Bees and Other Pollinators

In recent times, environmental groups have put pressure on the Environmental Protection Agency to ban neonicotinoids, a type of insecticide. In May 2015, the Obama Administration released a strategy called National Strategy to Promote the Health of Honey Bees and Other Pollinators. The administration announced it would include input from the pesticide industry in putting together the initiative.

The task force goal is "tackling and reducing the impact of multiple stressors on pollinator health, including pests and pathogens, reduced habitat, lack of nutritional resources, and exposure to pesticides."

The EPA and U.S. Department of Agriculture are leading the task force.

The Structure of Plant-pollinator Networks

Wild pollinators often visit a large number of plant species and plants are visited by a large number of pollinator species. All these relations together form a network of interactions between plants and pollinators. Surprising similarities were found in the structure of networks consisting out of the interactions between plants and pollinators. This structure was found to be similar in very different ecosystems on different continents, consisting of entirely different species.

The structure of plant-pollinator networks may have large consequences for the way in which pollinator communities respond to increasingly harsh conditions. Mathematical models, examining the consequences of this network structure for the stability of pollinator communities suggest that the specific way in which plant-pollinator networks are organized minimizes competition between pollinators and may even lead to strong indirect facilitation between pollinators when conditions are harsh. This allows pollinator species to survive together under harsh conditions. But it also means that pollinator species collapse simultaneously when conditions pass a critical point. This simultaneous collapse occurs, because pollinator species depend on each other when surviving under difficult conditions.

Such a community-wide collapse, involving many pollinator species, can occur suddenly when increasingly harsh conditions pass a critical point and recovery from such a collapse might not be easy. The improvement in conditions needed for pollinators to recover, could be substantially larger than the improvement needed to return to conditions at which the pollinator community collapsed.

Pollination Trap

Pollination traps or trap-flowers are plant flower structures that aid the trapping of insects, mainly flies, so as to enhance their effectiveness in pollination. The structures of pollination traps can include deep tubular corollas with downward pointing hairs, slippery surfaces, adhesive liquid, attractants (often deceiving the insects by the use of sexual attractants rather than nectar reward and therefore termed as deceptive pollination), flower closing and other mechanisms.

Arum with trap chamber at base

In many species of orchids, the flowers produce chemicals that deceive male insects by producing attractants that mimic their females. The males are then led into structures that ensure the transfer of pollen to the surfaces of the insects. Orchids in the genus *Pterostylis* have been found to attract male fungus gnats with chemical attractants and then trap them using a mobile petal lip. The general observation of insects being trapped and aiding pollination were made as early as 1872 by Thomas Frederic Cheeseman and did not go unnoticed by Charles Darwin who examined the adaptations of orchids for pollination. Slipper orchids have smooth landing surfaces that allow insects to slide into a container from which a window of light leads the insect outwards through a narrow passage where the pollen transfer occurs. The structures found in large flowers such as those of *Rafflesia* and some *Aristolochia* are also evolved to attract and trap pollinators.

Ceropegia rhynchantha, another trap flower

Trap-flowers that produce deceptive sexual chemicals to attract insects may often lack nectar rewards. Many fly-trapping flowers produce the smell of carrion.

Many members of the *Arum* family trap pollinators and the specific mechanisms vary with the insects involved.

Plants in the genus *Ceropegia* attract pollinating small flies (usually female) in a wide range of families, including Milichiidae, Chloropidae, Drosophilidae, Calliphoridae, Ephydridae, Sciaridae, Tachinidae, Scatopsidae, Phoridae, and Ceratopogonidae, and the pollinaria always attach to their probosces. An analysis of the scents emitted by *Ceropegia dolichophylla* showed the presence of spiroacetals which are rare in plants and common among insects. Milichid flies, which are kleptoparasites of arthropod predators, are attracted by these chemicals and become the pollinators of these plants.

Pollen Tube

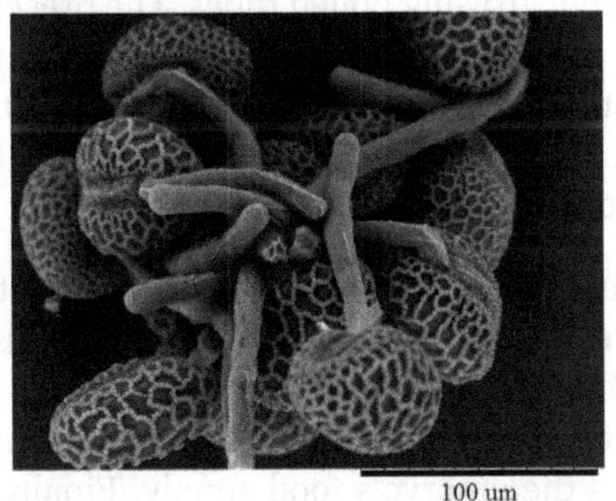

SEM image of pollen tubes growing from *Lily* pollen grains.

A pollen tube is part of the male gametophyte of seed plants. It acts as a conduit to transport the male gamete cells from the pollen grain, either from the stigma (in flowering plants) to the ovules at the base of the pistil, or directly through ovule tissue in some gymnosperms. In maize, this single cell can grow longer than 12 inches to traverse the length of the pistil.

Pollen tubes were first discovered by Giovanni Battista Amici.

The Pollen Tube in Angiosperms

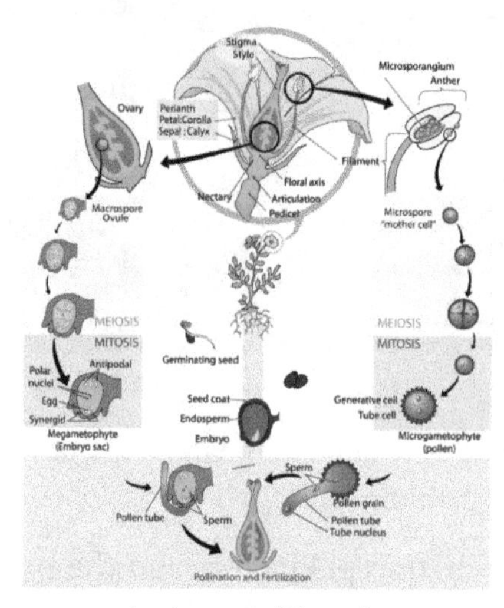

Angiosperm life cycle.

Angiosperm reproduction is a complex process that includes several steps that may vary among species. Each step is a vast procedure in its own right. Pollen is produced by the stamen, the male reproductive organ of the flower. Each pollen grain contains a vegetative cell, and a generative cell that divides to form two sperm cells. The pollen is delivered by the opening of anthers for subsequent pollination, that is, for the transfer of pollen grains to the pistil, the female reproductive organ. Pollination is usually carried out by wind, water or insects. The ovaries hold the ovules that produce the female gamete: the egg cell, which waits in place for fertilization.

Once a pollen grain settles on a compatible pistil, it may germinate in response to a sugary fluid secreted by the mature stigma of certain plants. Lipids at the surface of the stigma may also stimulate pollen tube growth for compatible pollen. Plants that are self-sterile often inhibit the pollen grains from their own flowers from growing pollen tubes. The presence of multiple grains of pollen has been observed to stimulate quicker pollen tube growth in some plants. The vegetative cell then produces the pollen tube, a tubular protrusion from the pollen grain, which carries the sperm cells within its cytoplasm. The sperm cells are the male gametes that will join with the egg cell and the central cell in double fertilization.

The germinated pollen tube must then drill its way through the nutrient-rich style and curl to the bottom of the ovary to reach the ovule. Once the pollen tube successfully attains an ovule, it delivers the two sperm cells with a burst. One of them fertilizes the egg cell to form an embryo, which will become the future plant. The other one fuses with both polar nuclei of the central cell to form the endosperm, which serves as the embryo's food supply. Finally, the ovary will develop into a fruit and the ovules will develop into seeds.

Pollen Tubes used to Study Cell Physiology

Pollen tubes are an excellent model for the understanding of plant cell behavior. They are easily cultivated in vitro and have a very dynamic cytoskeleton that polymerizes at very high rates, providing the pollen tube with interesting mechanical properties. The pollen tube has an unusual kind of growth; it extends exclusively at its apex. Extending the cell wall only at the tip minimizes friction between the tube and the invaded tissue. This tip growth is performed in a pulsating manner rather than in a steady fashion. Remarkably, the pollen tube's journey through the style often results in depth-to-diameter ratios above 100:1 and up to 1000:1 in certain species. The internal machinery and the external interactions that govern the dynamics of pollen tube growth are far from being fully understood.

Pollen Tube Guidance

Extensive work has been dedicated to comprehend how the pollen tube responds to extracellular guidance signals to achieve fertilization. It is believed that pollen tubes react to a combination of chemical, electrical, and mechanical cues during its journey through the pistil.Recently a group of scientists at Nagoya university,Japan has discovered the chemical responsible for pollen tube guidance to be AMOR (activated molecule for response-capability) which is a sugar molecule. Nevertheless, several aspects have already been identified as central in the process of pollen tube growth. The actin filaments in the cytoskeleton, the peculiar cell wall, secretory vesicle dynamics, and the flux of ions, to name a few, are some of the fundamental features readily identified as crucial, but whose role has not yet been completely elucidated.

Pollination Syndrome

Baltimore (*Euphydryas phaeton*) nectaring at daisy (*Argyranthemum*)

Pollination syndromes are suites of flower traits that have evolved in response to natural selection imposed by different pollen vectors, which can be abiotic (wind and water) or biotic, such as birds, bees, flies, and so forth. These traits include flower shape, size, colour, odour, reward type and amount, nectar composition, timing of flowering, etc. For example, tubular red flowers with copious nectar often attract birds; foul smelling flowers attract carrion flies or beetles, etc.

The "classical" pollination syndromes as they are currently defined were developed in the 19th century by the Italian botanist Federico Delpino. Although they have been useful in developing our understanding of plant-pollinator interactions, an uncritical acceptance of pollination syndromes as providing a framework for classifying these relationships is rather out of date.

Abiotic

These do not attract animal pollinators. Nevertheless, they often have suites of shared traits.

Plantago media, pollinated by wind or insects

Wind Pollination (Anemophily)

Flowers may be small and inconspicuous, as well as green and not showy. They produce enormous numbers of relatively small pollen grains (hence wind-pollinated plants may be allergens, but seldom are animal-pollinated plants allergenic). Their stigmas may be large and feathery to catch the pollen grains. Insects may visit them to collect pollen; in some cases, these are ineffective pollinators and exert little natural selection on the flowers, but there are also examples of ambophilous flowers which are both wind and insect pollinated. Anemophilous, or wind pollinated flowers, are usually small and inconspicuous, and do not possess a scent or produce nectar. The anthers may produce a large number of pollen grains, while the stamens are generally long and protrude out of flower.

Water Pollination (Hydrophily)

Water-pollinated plants are aquatic and pollen is released into the water. Water currents therefore act as a pollen vector in a similar way to wind currents. Their flowers tend to be small and inconspicuous with lots of pollen grains and large, feathery stigmas to catch the pollen. However, this is relatively uncommon (only 2% of pollination is hydrophily) and most aquatic plants are insect-pollinated, with flowers that emerge into the air. *Vallisneria* is an example.

Biotic

Sunflower pollinated by butterflies and bees

Bee Pollination (Melittophily)

Bee-pollinated flowers can be very variable in their size, shape and colouration. They can be open and bowl-shaped (radially symmetrical) or more complex and non-radially symmetric ("zygomorphic"), as is the case with many peas and foxgloves.

Some bee flowers tend to be yellow or blue, often with ultraviolet nectar guides and scent. Nectar, pollen, or both are offered as rewards in varying amounts. The sugar in the nectar tends to be sucrose-dominated. A few bees collect oil from special glands on the flower.

There are diverse types of bees (such as honeybees, bumblebees, and orchid bees), forming large groups that are quite distinctive in size, tongue length and behaviour (some solitary, some colonial); thus generalization about bees is difficult. Some plants can only be pollinated by bees because their anthers release pollen internally, and it must be shaken out by buzz pollination (also known as "sonication"). Bees are the only animals that perform this behaviour. Bumblebees sonicate, but honeybees do not.

Wasp Pollination

Wasps are also responsible for the pollination of several plants species, being important pollen vectors, and in some cases, even more efficient pollinators than bees.

Butterfly Pollination (Psychophily)

Hesperoyucca whipplei (moth-pollinated)

Butterfly-pollinated flowers tend to be large and showy, pink or lavender in colour, frequently have a landing area, and are usually scented. Since butterflies do not digest pollen (with one exception), more nectar is offered than pollen. The flowers have simple nectar guides with the nectaries usually hidden in narrow tubes or spurs, reached by the long tongue of the butterflies.

Moth Pollination (Phalaenophily)

Day-flying sphinx moth nectaring on Brazilian vervain

Among the more important moth pollinators are the hawk moths (Sphingidae). Their behaviour is similar to hummingbirds: they hover in front of flowers with rapid wingbeats. Most are nocturnal or crepuscular. So moth-pollinated flowers tend to be white, night-opening, large and showy with tubular corollas and a strong, sweet scent produced in the evening, night or early morning. A lot of nectar is produced to fuel the high metabolic rates needed to power their flight.

Other moths (Noctuids, Geometrids, Pyralids, for example) fly slowly and settle on the flower. They do not require as much nectar as the fast-flying hawk moths, and the flowers tend to be small (though they may be aggregated in heads).

Sapromyophilous *Stapelia gigantea*

Fly Pollination (Myophily and Sapromyophily)

Flies tend to be important pollinators in high-altitude and high-latitude systems, where they are numerous and other insect groups may be lacking. There are two main types of fly pollination: myophily and sapromyophily.

Myophily includes flies that feed on nectar and pollen as adults - particularly bee flies (Bombyliidae), hoverflies (Syrphidae), and others - and these regularly visit flowers. In contrast, male fruit flies (Tephritidae) are enticed by specific floral attractants emitted by some wild orchids which do not produce nectar. Chemicals emitted by the orchid act as the fly's sex pheromone precursor or booster. Myophilous plants tend not to emit a strong scent, are typically purple, violet, blue, and white, and have open dishes or tubes.

Sapromyophiles, on the other hand, normally visit dead animals or dung. They are attracted to flowers which mimic the odor of such objects. The plant provides them with no reward and they leave quickly unless it has traps to slow them down. Such plants are far less common than myophilous ones.

Bird Pollination (Ornithophily)

Although hummingbirds are the most familiar nectar-feeding birds for North Americans, there are analogous species in other parts of the world: sunbirds, honeyeaters, flowerpeckers, honeycreepers, bananaquits, flowerpiercers, lories and lorikeets. Hummingbirds are the oldest group, with the greatest degree of specialization on nectar. Flowers attractive to hummingbirds that can hover in front of the flower tend to be large red or orange tubes with a lot of dilute nectar, secreted during the day. Since birds do not have a strong response to scent, they tend to be odorless. Perching birds need a substantial landing platform, so sunbirds, honeyeaters, and the like are less associated with tubular flowers.

African baobab (bat-pollinated)

Bat Pollination (Chiropterophily)

Bat-pollinated flowers tend to be large and showy, white or light coloured, open at night and have strong odours. They are often large and bell-shaped. Bats drink the nectar, and these plants typically offer nectar for extended periods of time. Sight, smell, and echo-location are used to initially find the flowers, and excellent spatial memory is used to visit them repeatedly. In fact, bats can identify nectar-producing flowers using echolocation. In the New World, bat pollinated flowers often have sulfur-scented compounds, but this does not carry to other parts of the world. Bat-pollinated plants have bigger pollen than their relatives.

Beetle Pollination (Cantharophily)

Beetle-pollinated flowers are usually large, greenish or off-white in color and heavily scented. Scents may be spicy, fruity, or similar to decaying organic material. Most beetle-pollinated flowers are flattened or dish shaped, with pollen easily accessible, although they may include traps to keep the beetle longer. The plant's ovaries are usually well protected from the biting mouthparts of their pollinators. Beetles may be particularly important in some parts of the world such as semi-arid areas of southern Africa and southern California and the montane grasslands of KwaZulu-Natal in South Africa.

Biology

Pollination syndromes reflect convergent evolution towards forms (phenotypes) that limit the number of species of pollinators visiting the plant. They increase the functional specialization of the plant with regard to pollination, though this may not affect the ecological specialization (i.e. the number of species of pollinators within that functional group). They are responses to common selection pressures exerted by shared pollinators or abiotic pollen vectors, which generate correlations among traits. That is, if two distantly related plant species are both pollinated by nocturnal moths, for example, their flowers will converge on a form which is recognised by the moths (e.g. pale colour, sweet scent, nectar released at the base of a long tube, night-flowering).

Advantages of Specialization

- Efficiency of pollination: the rewards given to pollinators (commonly nectar or pollen or both, but sometimes oil, scents, resins, or wax) may be costly to produce. Nectar can be cheap, but pollen is generally expensive as it is relatively high in nitrogen compounds. Plants have evolved to obtain the maximum pollen transfer for the minimum reward delivered. Different pollinators, because of their size, shape, or behaviour, have different efficiencies of transfer of pollen. And the floral traits affect efficiency of transfer: columbine flowers were experimentally altered and presented to hawkmoths, and flower orientation, shape, and colour were found to affect visitation rates or pollen removal.
- Pollinator constancy: to efficiently transfer pollen, it is best for the plant if the pollinator focuses on one species of plant, ignoring other species. Otherwise, pollen may be dropped uselessly on the stigmas of other species. Animals, of course, do not aim to pollinate, they aim to collect food as fast as they can. However, many pollinator species exhibit constancy, passing up available flowers to focus on one plant species. Why should animals specialize on a plant species, rather than move to the next flower of any species? Although pollinator constancy was recognized by Aristotle, the benefits to animals are not yet fully understood. The most common hypothesis is that pollinators must learn to handle particular types of flowers, and they have limited capacity to learn different types. They can only efficiently gather rewards from one type of flower.

These honeybees selectively visit flowers from only one species for a period of time, as can be seen by the colour of the pollen in their baskets:

Advantages of Generalization

Pollinators fluctuate in abundance and activity independently of their plants, and any one species may fail to pollinate a plant in a particular year. Thus a plant may be at an advantage if it attracts several species or types of pollinators, ensuring pollen transfer every year. Many species of plants have the back-up option of self-pollination, if they are not self-incompatible.

Criticisms of the Syndromes

Whilst it is clear that pollination syndromes can be observed in nature, there has been much debate amongst scientists as to how frequent they are and to what extent we can use the classical syndromes to classify plant-pollinator interactions. Although some species of plants are visited only by one type of animal (i.e. they are functionally specialized), many plant species are visited by very different pollinators. For example, a flower may be pollinated by bees, butterflies, and birds. Strict specialization of plants relying on one species of pollinator is relatively rare, probably because it can result in variable reproductive success across years as pollinator populations vary significantly. In such cases, plants should generalize on a wide range of pollinators, and such ecological generalization is frequently found in nature. A study in Tasmania found the syndromes did not usefully predict the pollinators.

This debate has led to a critical re-evaluation of the syndromes, which suggests that on average about one third of the flowering plants can be classified into the classical syndromes. This reflects the fact that nature is much less predictable and straightforward than 19th Century biologists originally thought. Pollination syndromes can be thought of as extremes of a continuum of greater or lesser specialization or generalization onto particular functional groups of pollinators that exert similar selective pressures" and the frequency with which flowers conform to the expectations of the pollination syndromes is relatively rare. In addition, new types of plant-pollinator interaction, involving "unusual" pollinating animals are regularly being discovered, such as specialized pollination by spider hunting wasps (Pompilidae) and fruit chafers (Cetoniidae) in the eastern grasslands of South Africa. These plants do not fit into the classical syndromes, though they may show evidence of convergent evolution in their own right.

An analysis of flower traits and visitation in 49 species in the plant genus *Penstemon* found that it was possible to separate bird- and bee- pollinated species quite well, but only by using floral traits which were not considered in the classical accounts of syndromes, such as the details of anther opening. Although a recent review concluded that there is "overwhelming evidence that functional groups exert different selection pressures on floral traits", the sheer complexity and subtlety of plant-pollinator interactions (and the growing recognition that non-pollinating organisms such as seed predators can affect the evolution of flower traits) means that this debate is likely to continue for some time.

Seed Dispersal Syndrome

A seed dispersal syndrome is a mutualistic plant-animal interaction. Seed dispersal syndromes are morphological characters of seeds correlated to particular seed dispersal agents. Dispersal is the event by which individuals move from the site of their parents to establish in a new area. A seed disperser is the vector by which a seed moves from its parent to the resting place where the individual will establish, for instance an animal. Similar to the term syndrome, a diaspore is a morphological functional unit of a seed for dispersal purposes.

Characteristics for seed dispersal syndromes are commonly fruit colour, mass, and persistence. These syndrome characteristics are often associated with the fruit that carries the seeds. Fruits are packages for seeds, composed of nutritious tissues to feed animals. However, fruit pulp is not commonly used as a seed dispersal syndrome because pulp nutritional value does not enhance seed dispersal success. Animals interact with these fruits because they are a common food source for them. Although, not all seed dispersal syndromes have fruits because not all seeds are dispersed by animals. Suitable biological and environmental conditions of dispersal syndromes are needed for seed dispersal and invasion success such as temperature and moisture.

Seed dispersal syndromes are parallel to pollination syndromes, which are defined as floral characteristics that attract organisms as pollinators. They are considered parallels because they are both plant-animal interactions, which increase the reproductive success of a plant. However, seed dispersal syndromes are more common in gymnosperms, while pollination syndromes are found in angiosperms. Seeds disperse to increase the reproductive success of the plant. The farther away a seed is from a parent, the better its chances of survival and germination. Therefore, a plant should select certain traits to increase dispersal by a vector (i.e. bird) to increase the reproductive success of the plant.

Evolution

Seeds have evolved traits to reward animals to enhance their dispersal abilities. Differing foraging behaviours of animals can lead to selection of dispersal traits and spatial variation such as increase in seed size for mammal dispersal, which can limit seed production. Seed production is limited by some seed syndromes because of their cost to the plant. Therefore, seed dispersal syndromes will evolve in a plant when the trait benefit outweighs the cost. The seed dispersers themselves play an essential role in syndrome evolution. For example, birds put strong selection pressure on seeds for colour of fruits because of their enhanced vision. Illustrations of such colour evolution include green colour being produced because its photosynthesis abilities are less costly while red colour emerges as a byproduct for protection from arthropods.

For visible characteristic differences to develop between dispersers and non-dispersers a few conditions need to be met 1. Specialization must increase dispersal success whether morphological, physiological or behavioural 2. Energy investment for dispersal will be taken from energy investment of other traits 3. Dispersal traits will benefit the dispersers over non-dispersers. Phenotypic (visible characteristics) differences in non-dispersers and dispersers can be caused by external factors, kin competition, intraspecific competition and habitat quality.

History

In 1930, Ridley wrote an important book called The dispersal of plants throughout the world, which goes into detail about each form of dispersal; dispersal by wind, water, animals, birds, reptiles and fish, adhesion, and people. He details the morphology and traits for each dispersal method, which are later described as seed dispersal syndromes. This began the idea of seed trait selection being associated with a form of seed dispersal. Then in 1969 van der Pijl identified seed dispersal syndromes based on each mechanism of seed dispersal in his book Principles of Dispersal in Higher Plants. He is the pinnacle of seed dispersal syndromes and is cited by many scientists who study seed dispersal syndromes. He describes the morphology of interactions between fruits and flowers, and classifies dispersal in invertebrates, fish, reptiles, birds, mammals, ants, wind, water and the plant itself. Janson in 1983 continued the study on seed dispersal syndromes and classified seed dispersal syndromes of fruit by size, colour and husk or no husks in species of Peruvian tropical forest. He went in depth about the interaction between plants that have adapted to seed dispersal by birds and mammals. Willson, Irvine & Walsh in 1989 added more factors to the study of seed dispersal syndromes and looked at differing fleshy fruits and their correlation to moisture and differing ecological factors. They looked at bird-dispersal and mammal-dispersal and how the fruits differed in dispersal syndromes such as colour and size. These scientists began the theory and ideas behind seed dispersal syndromes that are crucial to the evolution of reproduction in plants.

Types and Functions

Dispersal syndromes have been previously classified by: size, colour, weight, protection, flesh type, number of seeds, weight and start time of ripening. Syndromes are often associated with the type of dispersal and morphology. Also chemical composition can influence the disperser's fruit choice. The following are types of seed dispersal and their syndromes.

Anemochory

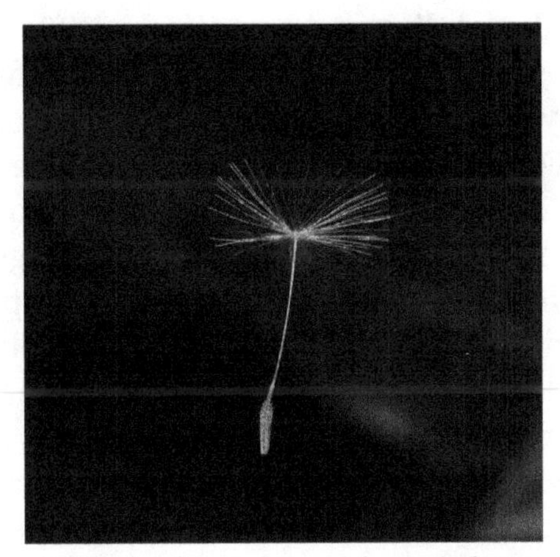

Example of a syndrome of anemochory

Anemochory is defined as seed dispersal by wind. Common dispersal syndromes of anemochory are wing structures and brown or dull coloured seeds without further rewards. Van der Pijl named seeds for anemochory flyers, rollers, or throwers to represent the seed dispersal syndromes and their behaviour. Flyers are typically categorized as dust diaspores, balloons, plumed or winged. Dust diaspores are small flat structures on seeds that appear to be the transition to wing diaspores, balloons are inflated seed characteristics and plumes are hairs or elongation seed characteristics. Wings have evolved to increase dispersal distance to promote gene flow. Anemochory is commonly found in open habitats, canopy trees, and dry season deciduous forests. Wind dispersers mature in the dry season for optimum high long-distance dispersal to increase success of germination.

Barochory

Barochory is seed dispersal by gravity alone in which a plant's seeds fall beneath the parent plant. These seeds commonly have heavy seed dispersal syndromes. However, heavy seeds may not be a form of seed dispersal syndrome, but a random seed characteristic that has no dispersal purpose. It has been thought that barochory does not develop a seed dispersal syndrome because it does not select for characters to enhance dispersal. It is questionable whether barochory is dispersal at all.

Hydrochory

Hydrochory is seed dispersal by water. Seeds can disperse by rain or ice or be submerged in water. Seeds dispersed by water need to have the ability to float and resist water damage. They often have hairs to assist with enlargement and floating. More features that cause floating are air space, lightweight tissues and corky tissues. Hydrochory syndromes are most common in aquatic plants.

Zoochory

Zoochory is the dispersal of seeds by animals and can be further divided into three classes.

1. Endozoochory is seed dispersal inside animals,
2. Synzoochory is dispersal of diaspores by the mouthparts of animals, and
3. Epizoochory is the accidental dispersal by animals. Differing characteristics of zoochory syndromes include coloured fruits, scented fruits, and different textures for different animals. Endozoochory syndrome characteristics will develop based on palatability of the fruit by an organism. For example, mammals are attracted to scent of a seed and birds are attracted to colour. Endozoochory syndromes have evolved to be ingested by animals and later bypassed in a new environment so the seed can germinate. Synzoochory should possess hard skins to protect seeds from damage of mouthparts; for example, sharp beaks on animals such as birds or turtles. Epizoochory commonly has burrs or spines to transport seeds on the outside of animals. These syndromes are highly associated with animals that have fur, while burrs would be lacking on seeds that are dispersed by reptiles because of their smooth skin. It is believed that not all animals that interact with plant fruits are dispersers because some animals do not increase the successful dispersal of seeds but consume and destroy them. Therefore, some animals are dispersers and some are consumers.

Mammalochory

Mammalochory is specifically the seed dispersal by mammals. The dispersal syndromes for mammalochory include large fleshy fruit, green or dull coloured fruits, and husked or unhusked. The seeds tend to have more protection to prevent mechanical destruction. Mammals rely on smell more than vision for foraging, which causes the seeds they disperse to be more scented compared to bird-dispersed seeds. Animal-dispersed seeds ripen in rainy season when foraging activity is high, resulting in fleshy diaspores. Mammals consume fruits whole or in smaller pieces, which explains the larger seed syndromes. Mammalochory syndromes can increase the reproductive success of the plant compared to seed dispersal syndromes of a plant associated with barochory for example. An example of seed dispersal syndromes associated with mammals that increases reproductive success would be seed-consuming rodents that increase germination by burial of seeds.

Ornithochory

Ornithochory is seed dispersal by birds. Common syndrome characteristics include small fleshy fruits with bright colours and without husks. Ornithochory is common in temperate zones and oceanic islands because of absence of native mammals. Birds have heightened colour vision and swallow seeds and fruits whole, explaining the small and coloured characteristics of dispersal syndromes. Birds have a weak sense of smell, therefore ornithochory syndromes would specialize more in colour than scent, in comparison to mammalochory. Ornithochory can increase the reproductive success of a plant because a bird's digestive tract increases seed germination after it has been bypassed and dispersed by the bird.

Myrmecochory

Myrmecochory is seed dispersal by ants. Myrmecochory is considered an ant-plant mutualistic relationship. The common syndrome traits for myrmecochory are elaisomes, and are often hard and difficult to damage. Elaisomes are structures that attract ants because they are high in lipid content, providing important nutrients for the ant. Without ants, seed dispersal becomes barochory and dispersal success declines. It is debated if ants are good dispersers and if plants would select for ant dispersal. Ants do clearly interact with seeds, however ants cannot travel very long distances. Therefore, would a plant select for an ant over a bird when birds can disperse seeds much farther than ants, increasing a plant's reproductive success.

Problems in Seed Dispersal Syndromes

Many scientists are skeptical whether seed dispersal syndromes actually exist because their parallel, pollination syndromes, are often disputed in scientific literature. Seed dispersal syndromes do not have much disagreement among scientists. Whether this is due to lack of research or interest in seed dispersal syndromes, or that scientists agree with the idea of seed dispersal syndromes. It also may be that seed dispersal syndromes are harder to test because once seeds disperse they are difficult to collect and study. Jordano (1995) states that the evolution of fruit traits for seed dispersal success is only dependent on diameter. This is one scientist's perspective but does not appear to be the common consensus among scientists. Colour and olfaction are other common seed dispersal syndromes tested and discussed in scientific literature. One limitation to seed dispersal syndromes mentioned is the limited definitions of syndrome characteristics such as odour or texture. It is possible that there has not been enough research to test these characteristics or they do not play a role in seed dispersal syndromes.

The differences in seed dispersal syndromes appear to be weak, but do exist. There needs to be consideration for the possibility that these syndromes evolved not to benefit seed dispersal but possibility to combat other selective pressures. For example, syndromes may have developed to combat predation or environmental hazards. Predation could produce a secondary metabolite syndrome. Secondary metabolites are compounds that are not used for the primary function of a plant and are normally used as defense mechanisms.

Further Research

Seed dispersal syndromes have not been studied in complete breadth for every seed dispersal method. Therefore, further research should be conducted to fill the gaps of knowledge about dispersal syndromes. The following are problems areas or directions research can continue on the study of seed dispersal syndromes. There is a lack of understanding of morphology in correlation to behavioural traits of dispersers. Research in this area would assist in the understanding of why particular dispersers are selected by plants to enhance reproductive success. Also, understanding movement strategies of factors affecting departure to settlement is important in determining whether seed dispersal syndromes are only affect by plant selection for a disperser. There are few studies concerning phenotype-dependent dispersal and how it affects spatial structures of populations. Distance of dispersal is not researched in enough detail to correlate to a seed dispersal syndrome. More experimental field studies on plant-animal interactions regarding seed dispersal need to be conducted for a thorough understanding of seed dispersal syndromes. There is limited knowledge about the presence of elaisomes and ant behaviour affecting seed dispersal, and how ant-plant interactions evolved under various plant traits. Understanding these interactions would help clarify if myrmecochory did evolve seed dispersal syndromes. Micro and macroevolutionary processes are needed to determine the effects of biological dispersal of seeds. There cannot be inferences about seed dispersal syndromes without robust phylogenies and evolutionary studies. There is also a gap in the understanding of genetic consequences of zoochory. Using genetics could help clarify if these syndromes were formed at random or if they correspond to evolution of seed dispersal. It is unclear if these seed dispersal syndromes evolved for specialization between plants and animals to increase seed dispersal success or if these syndromes are simply formed from generalist plant-animal interactions. Understanding these relationships would clarify the confusion about seed dispersal syndromes and if they are true examples of evolution increasing plant reproductive success or if they have developed without selective pressures.

Ornithophily

Hummingbird *Phaethornis longirostris* on an *Etlingera* inflorescence

Ornithophily or bird pollination is the pollination of flowering plants by birds. This coevolutionary association is derived from insect pollination (entomophily) and is particularly well developed in some parts of the world, especially in the tropics and on some island chains. The association involves several distinctive plant adaptations forming a "pollination syndrome". The plants typically have colourful, often red, flowers with long tubular structures holding ample nectar and orientations of the stamen and stigma that ensure contact with the pollinator. Birds involved in ornithophily tend to be specialist nectarivores with brushy tongues, long bills, capable of hovering flight or are light enough to perch on the flower structures.

Plant Adaptations

A lesser violetear

Inflorescences of *Butea* allow birds to perch on the stalk

Bird pollination is considered as a costly strategy for plants and it evolves only where there are particular benefits for the plant. High altitude ecosystems that lack insect pollinators, those in dry regions or isolated islands tend to favour the evolution of ornithophily in plants.

Plants adaptations can be grouped into mechanisms that attract birds, those that exclude insects, protect against nectar theft and pollination mechanisms in the strict sense. The ovules of bird flowers also tend to have adaptations that protect them from damage.

Most bird pollinated flowers are red and have a lot of nectar. They also tend to be unscented. Flowers with generalist pollinators tend to have dilute nectar but those that have specialist pollinators such as hummingbirds or sunbirds tend to have more concentrated nectar. The nectar of ornithophilous flowers vary in the sugar composition, with hexoses being high in passerine pollinated species while those that are insect pollinated tend to be sucrose rich. Hummingbird pollinated flowers however tend to be sucrose rich. Many plants of the family Loranthaceae have explosive flowers that shower pollen on a bird that forages near it. They are associated mainly with flowerpeckers in the Dicaeidae family. In Australia, some species of *Banksia* have flowers that open in response to bird actions thereby reducing the wastage of pollen. In tropical dry forests in southern India, ornithophilous flowers were found to bloom mainly in the hot dry season. As many as 129 species of North American plants have evolved ornithophilous associations. Nearly a fourth of the 900 species of the genus *Salvia* are bird pollinated in the South African region. Tropical China and the adjacent Indochinese countries harbor relatively few bird-pollinated flowers, among them is *Rhodoleia championii*, a member of the Hamamelidaceae family, which at any one site can be visited and pollinated by up to seven species of nectar-foraging birds, including Japanese white-eyes (*Zosterops japonicus*, Zosteropidae) and fork-tailed sunbirds (Aethopyga christinae, Nectariniidae).

The rat's tail babiana (*Babiana ringens*) is a species of plant that produces a strong stalk within the inflorescence that serves as a perch for the malachite sunbird as it visits the flower. *Heliconias* have special sticky threads that help in the adhesion of pollen to smooth structures such as the bill of a hummingbird. Some African orchids of the genus *Disa* have pollinaria that stick to the feet of visiting sunbirds.

Plants need to protect against nectar being taken by non-pollinators. These agents are classified into nectar robbers, which may destroy the flower, for example cut the flower at the base to obtain nectar and nectar thieves that obtain nectar without pollinating the flower.

Bird Adaptations

Ruby-throated hummingbird (*Archilochus colubris*) at scarlet beebalm flowers (*Monarda didyma*)

The main families of specialized nectar feeding birds that are involved in ornithophily are the hummingbirds (Trochilidae), sunbirds (Nectariniidae), and the honey-eaters (Meliphagidae). Other important bird groups include those in the families the Icteridae, the honeycreepers (Thraupidae, Drepanidae), white-eyes (Zosteropidae) and the South African sugar-birds (Promeropidae). Birds may obtain nectar either by perching or by hovering with the latter mainly found in the hummingbirds and sunbirds. Within the hummingbirds, two kinds of foraging are noted with territorial "hermit" hummingbirds and the non-hermits which forage longer distances

Hummingbirds have the ability to digest sucrose unlike many passerines that prefer hexoses (fructose and glucose). Starlings and their relatives will completely avoid sucrose. Nectar feeding birds typically have a mechanism to quickly excrete excess water. They may have to drink four to five times their body mass of liquid during the day to obtain enough energy. Hummingbirds are capable of excreting nitrogenous wastes as ammonia since they can afford more water loss than birds that feed on low-moisture food sources. Hummingbirds and sunbirds also have special anatomical and physiological adaptations that allow them to quickly excrete excess water. Hummingbirds are also able to turn off their kidney function at night.

In some birds such as white-eyes, the pollen dusted by the plants on the forehead of the birds may increase the wear of these feathers leading to increased moulting and replacement.

Other Associations

Several mite species (mainly in the genera *Proctolaelaps*, *Tropicoseius* and *Rhinoseius*, family Ascidae) have evolved a phoretic mode of life, climbing into the nostrils of hummingbirds that visit flowers and hitching a ride to other flowers where they can feed on the nectar. Hummingbird flower mites favour plants in the families of Heliconiaceae, Costaceae, Zingiberaceae, Amaryllidaceae, Rubiaceae, Apocynaceae, Bromeliaceae, Gesneriaceae, Lobeliaceae and Ericaceae, members of which are associated with hummingbirds.

Buzz Pollination

A bee using buzz pollination

Buzz pollination or sonication is a technique used by some bees, such as the *Bombus morio* and many other bumble bees, to release pollen which is more or less firmly held by the anthers. The anther of buzz-pollinated species of plants is typically tubular, with an opening at only one end, and the pollen inside is smooth-grained and firmly attached. With self-fertile plants such as tomatoes, wind may be sufficient to shake loose the pollen through pores in the anther and accomplish pollination. Visits by bees may also shake loose some pollen, but more efficient pollination of those plants is accomplished by a few insect species who specialize in *sonication* or buzz pollination.

In order to release the pollen, bumblebees and some species of solitary bees are able to grab onto the flower and move their flight muscles rapidly, causing the flower and anthers to vibrate, dislodging pollen. This resonant vibration is called *buzz pollination*. The honeybee cannot perform buzz pollination. About 8% of the flowers of the world are primarily pollinated using buzz pollination.

Plants Pollinated by Buzz Pollination

The following plants are pollinated more efficiently by buzz pollination:

- All *Dodecatheon* (shooting stars)
- *Heliamphora*
- Many members of the Solanaceae family

- Many species of the genus *Solanum*
    - Eggplants
    - Potatoes
    - Tomatoes
    - *Solanum cinereum*, an Australian shrub
- *Hibbertia*
- *Dianella* (flax lilies)
- Some members of the genus *Vaccinium*
    - Blueberries
    - Cranberries
- *Arctostaphylos* – manzanita
- Some Fabaceae
    - Senna

Techniques for Agricultural Pollination of Species Normally Requiring Buzz Pollination

Greenhouse grown tomatoes are unproductive without aid in pollination. Traditionally, pollination has been done by shaking using electric vibrators (one brand name was "Electric Bee"), however, it has been found to be less expensive in human labor and plant breakage to use bumblebees within the greenhouses. In Australia, as bumblebees are not native, and Australia has a number of widely publicised environmental disasters caused by escaped introduced species ("feral species"), research is under way to adapt the use of the Australian native *Amegilla cingulata* (blue banded bees) for the same task. This research is, however, competing with lobbying by potential importers of bumblebees, who would rather use those, disregarding the risk and the potential for developing a "home grown" solution.

Fruit Tree Pollination

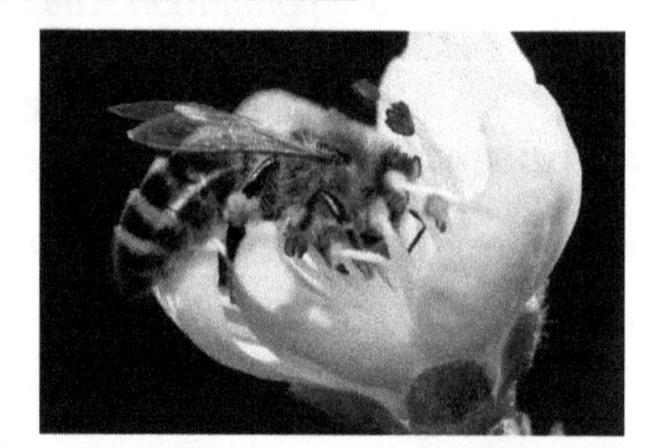

A European honey bee pollinates a peach flower while collecting nectar.

Pollination of fruit trees is required to produce seeds with surrounding fruit. It is the process of moving pollen from the anther to the stigma, either in the same flower or in another flower. Some tree species, including many fruit trees, do not produce fruit from self-pollination, so pollinizer trees are planted in orchards.

The pollination process requires a carrier for the pollen, which can be animal, wind, or human intervention (by hand-pollination or by using a pollen sprayer). Cross pollination produces seeds with a different genetic makeup from the parent plants; such seeds may be created deliberately as part of a selective breeding program for fruit trees with desired attributes. Trees that are cross-pollinated or pollinated via an insect pollinator produce more fruit than trees with flowers that just self-pollinate. In fruit trees, bees are an essential part of the pollination process for the formation of fruit.

Pollination of fruit trees around the world has been highly studied for hundreds of years. There is a lot of information known about fruit tree pollination from temperate climates, but very little information known about fruit tree pollination from tropical climates. Fruits from temperature climates include apples, pears, plums, peaches, cherries, berries, grapes, and nuts which are considered dry fruits. Fruits from tropical climates include bananas, pineapples, papayas, passion fruit, avocado, mango, and members of the genus *Citrus*.

Apple

Most apples are self-incompatible, that is, they do not produce fruit when pollinated from a flower of the same tree or from another tree of the same cultivar, and must be cross pollinated. A few are described as "self-fertile" and are capable of self-pollination, although even those tend to carry larger crops when cross pollinated from a suitable pollenizer. A relatively small number of cultivars are "triploid", meaning that they provide almost no viable pollen for themselves or other apple trees. Apples that can pollinate one another are grouped by the time they flower so cross-pollinators are in bloom at the same time. Pollination management is an important component of apple culture. Before planting, it is important to arrange for pollenizers - varieties of apple or crabapple that provide plentiful, viable and compatible pollen. Orchard blocks may alternate rows of compatible varieties, or may plant crabapple trees, or graft on limbs of crabapple. Some varieties produce very little pollen, or the pollen is sterile, so these are not good pollenizers. Good-quality nurseries have pollenizer compatibility lists. Growers with old orchard blocks of single varieties sometimes provide bouquets of crabapple blossoms in drums or pails in the orchard for pollenizers. Home growers with a single tree and no other variety in the neighborhood can do the same on a smaller scale.

During the bloom each season, commercial apple growers usually provide pollinators to carry the pollen. Honeybee hives are most commonly used in the United States, and arrangements may be made with a commercial beekeeper who supply hives for a fee. Honeybees of the genus *Apis* are the most common pollinator for apple trees, although members of the genera *Andrena*, *Bombus*, *Halictus*, and *Osmia* pollinate apple trees in the wild. Solitary bees such as ground-nesting mining bees (*Andrena*) may play a far bigger role in pollination than at one time suspected. Bumble bees are sometimes present in orchards, but not usually in enough quantity to be significant pollinators; in the home garden with only a few trees, their role may be much greater.

Increasingly Orchard bees (spring mason bees) are being used in fruit tree pollination. According to British writer Christopher O'Toole in his book *The Red Mason Bee*, *Osmia rufa* is a much more efficient pollinator of orchard crops (in Europe) than honey bees. Both *O. rufa* and *O. cornuta* are used in Europe, while in western North America, the "Blue Orchard Bee" (*Osmia lignaria*, more black than blue in color) is a proven orchard pollinator. In Japan, the Japanese Orchard Bee -- the hornfaced bee, *Osmia cornifrons* -- provides up to 80% of the apple pollination. Beyond Japan, *Osmia cornifrons* is also used increasingly in the eastern US, because like other mason bees it is up to 100 times more efficient than the honeybee -- a mere 600 hornfaced bees being required per hectare, as opposed to tens of thousands of honeybees. Home growers may find these more acceptable in suburban locations because they rarely sting.

Symptoms of inadequate pollination are small and misshapen apples, and slowness to ripen. The seeds can be counted to evaluate pollination. Well-pollinated apples have best quality, and will have seven to ten seeds. Apples with fewer than three seeds will usually not mature and will drop from the trees in the early summer. Inadequate pollination can result from either a lack of pollinators or pollenizers, or from poor pollinating weather at bloom time. Multiple bee visits are usually required to deliver sufficient grains of pollen to accomplish complete pollination.

Almonds

The blossoms of most California Almond varieties are self-incompatible, requiring cross-pollination with other varieties to produce a crop. The single most important factor determining a good yield is pollination during the bloom period. More than a million colonies of honey bees are placed in California Almond orchards at the beginning of the bloom period to pollinate the crop. California beekeepers alone cannot supply this critical need, which is why honey bees travel across the country to the San Joaqin Valley each year. Although the recommended number of hives per acre is 2 to 3, due to the high demand in conjunction with the reduced availability of commercial beehives, many almond growers have to make do with a lower hive density during pollination. These growers started using semiochemical formulations, like SPLAT Bloom, to compensate for the low hive density. SPLAT Bloom manipulates the behavior of the bees, inciting them to spend more time foraging, and thus pollinating flowers in entire the almond orchard (increasing pollination and fruit set), not only close to the hive.

Research into self-fertile almonds has led to the development of several almond varieties that do not need a pollinator tree. Working from an old Spanish variety, "Tuono", researchers in 2008 started making available new self-fertile varieties that more closely match the qualities of the popular "Nonpareil" almond -- including varieties "Lone Star" and "Independence". While self-fertile trees do not require pollen from a second tree variety for fruit / nut set, they still depend upon insect pollination for good production. Almond growers with self-fertile almonds report excellent nut set with half (or less) the number of bees in the field.

Pear

Like apples, pears are self-incompatible and need to attract insects in order to be pollinated and produce fruit. One notable difference from apples is that pear blossoms are much less attractive to bees due to their pale coloring and light odor. Bees may abandon the pear blossoms to visit dandelions or a nearby apple orchard. The majority of pollinators of pear trees are honey bees, although pears are also visited by blow flies and hoverflies. A way to combat the low attraction of honey bees to pear blossoms is to use bee attractants to entice the bees to pollinate the flowers. Bee attractants may include pheromones that mimic the brood pheromone or the juvenile pheromone, or other attractants. There are also other methods for attracting honey bees to pear blossoms. One is saturation pollination, that is to stock so many bees that all area blossoms are worked regardless of the attractiveness to the bees. Another method is to delay the movement of the beehives into the orchards until there is about 30 percent bloom. The bees are moved into the orchard during the night and will usually visit the pear blossoms for a few hours until they discover the richer nectar sources. The recommended number of hives per acre is one.

Citrus

Many citrus cultivars are seedless and are produced parthenocarpically without pollination. Some cultivars may be capable of producing fruit either way, having seeds in the segments, if pollinated, and no seeds if not. Citrus that requires pollination may be self-compatible, thus pollen must be moved only a short distance from the anther to the stigma by a pollinator. Some citrus, such as Meyer Lemons, are popular container plants. When these bloom indoors, they often suffer from blossom drop because no pollinators have access. Hand pollination by a human pollinator is a solution.

A few citrus, including some tangelos and tangerines, are self-incompatible, and require cross pollination. Pollinizers must be planned when groves are planted. Managed honeybee hives at bloom time are often used to ensure adequate pollination.

References

- Mauseth, James D. Botany: An Introduction to Plant Biology. Publisher: Jones & Bartlett, 2008 ISBN 978-0-7637-5345-0
- Raghavan, Valayamghat (1997). Molecular Embryology of Flowering Plants. Cambridge University Press. pp. 210–216. ISBN 978-0-521-55246-2.
- Campbell, Neil A.; Reece, Jane B. (2002). Biology (6th edition). Pearson Education. pp. 600–612. ISBN 978-0-201-75054-6.
- Cronk, J. K.; Fennessy, M. Siobhan (2001). Wetland plants: biology and ecology. Boca Raton, Fla.: Lewis Publishers. p. 166. ISBN 1-56670-372-7.
- Glover, Beverly J. (2007). Understanding flowers and flowering: an integrated approach. Oxford University Press. p. 127. ISBN 0-19-856596-8.
- Baskin, Carol C.; Baskin, Jerry M. (2001). Seeds: Ecology, Biogeography, and Evolution of Dormancy and Germination. Elsevier. p. 215. ISBN 978-0-12-080263-0.
- A. K. Shukla; M. R. Vijayaraghavan; Bharti Chaudhry (1998). "Abiotic pollination". Biology Of Pollen. APH Publishing. pp. 67–69. ISBN 9788170249245.
- Dave Moore (2001). "Insects of palm flowers and fruits". In F.W. Howard; D. Moore; R.M. Giblin-Davis; R.G. Abad. Insects on Palms. CAB International. pp. 233–266. ISBN 9780851997056.
- McGhee, George R. (2011). Convergent Evolution: Limited Forms Most Beautiful. MIT Press. pp. 118–120. ISBN 978-0-262-01642-1.
- Faegri, K.; Van Der Pijl, L. (2013). Principles of Pollination Ecology. Elsevier. pp. 34–36. ISBN 978-1-4832-9303-5.

- Prance, Ghillean T. (1996). The Earth Under Threat: A Christian Perspective. Wild Goose Publications. p. 14. ISBN 978-0-947988-80-7.
- Faegri, K.; Van Der Pijl, L. (2013). Principles of Pollination Ecology. Elsevier. pp. 102–110. ISBN 978-1-4832-9303-5.
- Faegri, K.; Van Der Pijl, L. (2013). Principles of Pollination Ecology. Elsevier. pp. 176–177. ISBN 978-1-4832-9303-5.

8

Plant Reproduction

The process of reproduction that takes place in plants is known as plant reproduction. Certation, diaspore, flower, gynoecium, petal, stamen and stigma are some of the aspects that have been explained in the section. The aspects elucidated in this section are of vital importance, and provide a better understanding of plant reproduction system.

Plant Reproduction

Plant reproduction is the production of new individuals or offspring in plants, which can be accomplished by sexual or asexual reproduction. Sexual reproduction produces offspring by the fusion of gametes, resulting in offspring genetically different from the parent or parents. Asexual reproduction produces new individuals without the fusion of gametes, genetically identical to the parent plants and each other, except when mutations occur. In seed plants, the offspring can be packaged in a protective seed, which is used as an agent of dispersal.

Bryophyllum, a plant that reproduces asexually via new shoots from the leaves

Asexual Reproduction

Plants have two main types of asexual reproduction in which new plants are produced that are genetically identical clones of the parent individual. Vegetative reproduction involves a vegetative piece of the original plant (budding, tillering, etc.) and is distinguished from *apomixis*, which is a replacement for sexual reproduction, and in some cases involves seeds. Apomixis occurs in many plant species and also in some non-plant organisms. For apomixis and similar processes in non-plant organisms.

Natural vegetative reproduction is mostly a process found in herbaceous and woody perennial plants, and typically involves structural modifications of the stem or roots and in a few species leaves. Most plant species that employ vegetative reproduction do so as a means to perennialize the plants, allowing them to survive from one season to the next and often facilitating their expansion in size. A plant that persists in a location through vegetative reproduction of individuals constitutes a clonal colony; a single ramet, or apparent individual, of a clonal colony is genetically identical to all others in the same colony. The distance that a plant can move during vegetative reproduction is limited, though some plants can produce ramets from branching rhizomes or stolons that cover a wide area, often in only a few growing seasons. In a sense, this process is not one of reproduction but one of survival and expansion of biomass of the individual. When an individual organism increases in size via cell multiplication and remains intact, the process is called vegetative growth. However, in vegetative reproduction, the new plants that result are new individuals in almost every respect except genetic. A major disadvantage to vegetative reproduction, is the transmission of pathogens from parent to offspring; it is uncommon for pathogens to be transmitted from the plant to its seeds (in sexual reproduction or in apomixis), though there are occasions when it occurs.

Seeds generated by apomixis are a means of asexual reproduction, involving the formation and dispersal of seeds that do not originate from the fertilization of the embryos. Hawkweed (*Hieracium*), dandelion (*Taraxacum*), some Citrus (*Citrus*) and Kentucky blue grass (*Poa pratensis*) all use this form of asexual reproduction. Pseudogamy occurs in some plants that have apomictic seeds, where pollination is often needed to initiate embryo growth, though the pollen contributes no genetic material to the developing offspring. Other forms of apomixis occur in plants also, including the generation of a plantlet in replacement of a seed or the generation of bulbils instead of flowers, where new cloned individuals are produced.

Structures

A rhizome is a modified underground stem serving as an organ of vegetative reproduction; the growing tips of the rhizome can separate as new plants, e.g., Polypody, Iris, Couch Grass and Nettles.

Prostrate aerial stems, called runners or stolons are important vegetative reproduction organs in some species, such as the strawberry, numerous grasses, and some ferns.

Adventitious buds form on roots near the ground surface, on damaged stems (as on the stumps of cut trees), or on old roots. These develop into above-ground stems and leaves. A form of budding called suckering is the reproduction or regeneration of a plant by shoots that arise from an existing root system. Species that characteristically produce suckers include Elm (*Ulmus*), Dandelion (*Taraxacum*), and many members of the Rose family such as *Rosa* and *Rubus*.

Plants like onion (*Allium cepa*), hyacinth (*Hyacinth*), narcissus (*Narcissus*) and tulips (*Tulipa*) reproduce by dividing their underground bulbs into more bulbs. Other plants like potatoes (*Solanum tuberosum*) and dahlia (*Dahlia*) reproduce by a similar method involving underground tubers. Gladioli and crocuses (*Crocus*) reproduce in a similar way with corms.

Usage

The most common form of plant reproduction utilized by people is seeds, but a number of asexual methods are utilized which are usually enhancements of natural processes, including: cutting, grafting, budding, layering, division, sectioning of rhizomes, roots, tubers, bulbs, stolons, tillers (suckers), etc., and artificial propagation by laboratory tissue cloning. Asexual methods are most often used to propagate cultivars with individual desirable characteristics that do not come true from seed. Fruit tree propagation is frequently performed by budding or grafting desirable cultivars (clones), onto rootstocks that are also clones, propagated by stooling.

In horticulture, a "cutting" is a branch that has been cut off from a mother plant below an internode and then rooted, often with the help of a rooting liquid or powder containing hormones. When a full root has formed and leaves begin to sprout anew, the clone is a self-sufficient plant, genetically identical to the mother plant. Examples include cuttings from the stems of blackberries (*Rubus occidentalis*), African violets (*Saintpaulia*), verbenas (*Verbena*) to produce new plants. A related use of cuttings is grafting, where a stem or bud is joined onto a different stem. Nurseries offer for sale trees with grafted stems that can produce four or more varieties of related fruits, including apples. The most common usage of grafting is the propagation of cultivars onto already rooted plants, sometimes the rootstock is used to dwarf the plants or protect them from root damaging pathogens.

Since vegetatively propagated plants are clones, they are important tools in plant research. When a clone is grown in various conditions, differences in growth can be ascribed to environmental effects instead of genetic differences.

Sexual Reproduction

Sexual reproduction involves two fundamental processes: meiosis, which rearranges the genes and reduces the number of chromosomes, and fertilization, which restores the chromosome to a complete diploid number. In between these two processes, different types of plants and algae vary, but many of them, including all land plants, undergo alternation of generations, with two different multicellular structures (phases), a gametophyte and a sporophyte. The evolutionary origin and adaptive significance of sexual reproduction are discussed in the pages "Evolution of sexual reproduction" and "Origin and function of meiosis."

The gametophyte is the multicellular structure (plant) that is haploid, containing a single set of chromosomes in each cell. The gametophyte produces male or female gametes (or both), by a process of cell division called mitosis. In vascular plants with separate gametophytes, female gametophytes are known as mega gametophytes (mega=large, they produce the large egg cells) and the male gametophytes are called micro gametophytes (micro=small, they produce the small sperm cells).

The fusion of male and female gametes (fertilization) produces a diploid zygote, which develops by mitotic cell divisions into a multicellular sporophyte.

The mature sporophyte produces spores by meiosis, sometimes referred to as "reduction division" because the chromosome pairs are separated once again to form single sets.

In mosses and liverworts the gametophyte is relatively large, and the sporophyte is a much smaller structure that is never separated from the gametophyte. In ferns, gymnosperms, and flowering plants (angiosperms), the gametophytes are relatively small and the sporophyte is much larger. In gymnosperms and flowering plants the mega gametophyte is contained within the ovule (that may develop into a seed) and the micro gametophyte is contained within a pollen grain.

History of Sexual Reproduction

Unlike animals, plants are immobile, and cannot seek out sexual partners for reproduction. In the evolution of early plants, abiotic means, including water and wind, transported sperm for reproduction. The first plants were aquatic, as described in the page "Evolutionary history of plants", and released sperm freely into the water to be carried with the currents. Primitive land plants like liverworts and mosses had motile sperm that swam in a thin film of water or were splashed in water droplets from the male reproduction organs onto the female organs. As taller and more complex plants evolved, modifications in the alternation of generations evolved; in the Paleozoic era progymnosperms reproduced by using spores dispersed on the wind. The seed plants including seed ferns, conifers and cordaites, which were all gymnosperms, evolved 350 million years ago; they had pollen grains that contained the male gametes for protection of the sperm during the process of transfer from the male to female parts. It is believed that insects fed on the pollen, and plants thus evolved to use insects to actively carry pollen from one plant to the next. Seed producing plants, which include the angiosperms and the gymnosperms, have heteromorphic alternation of generations with large sporophytes containing much reduced gametophytes. Angiosperms have distinctive reproductive organs called flowers, with carpels, and the female gametophyte is greatly reduced to a female embryo sac, with as few as eight cells. The male gametophyte consists of the pollen grains. The sperm of seed plants are non-motile, except for two older groups of plants, the Cycadophyta and the Ginkgophyta, which have flagellated sperm.

Flowering Plants

Flowering plants are the dominant plant form on land and they reproduce by sexual and asexual means. Often their most distinguishing feature is their reproductive organs, commonly called flowers. Sexual reproduction in flowering plants involves the production of male and female gametes, the transfer of the male gametes to the female ovules in a process called pollination. After pollination occurs, fertilization happens and the ovules grow into seeds within a fruit. After the seeds are ready for dispersal, the fruit ripens and by various means the seeds are freed from the fruit and after varying amounts of time and under specific conditions the seeds germinate and grow into the next generation.

The anther produces male gametophytes which are pollen grains, which attach to the stigma on top of a carpel, in which the female gametophytes (inside ovules) are located. After the pollen tube grows through the carpel's style, the sperm from the pollen grain migrate into the ovule to fertilize the egg cell and central cell within the female gametophyte in a process termed double fertilization. The resulting zygote develops into an embryo, while the triploid endosperm (one sperm cell plus a binucleate female cell) and female tissues of the ovule give rise to the surrounding tissues in the developing seed. The ovary, which produced the female gametophyte(s), then grows into a fruit, which surrounds the seed(s). Plants may either self-pollinate or cross-pollinate.

Pollination

In plants that use insects or other animals to move pollen from one flower to the next, plants have developed greatly modified flower parts to attract pollinators and to facilitate the movement of pollen from one flower to the insect and from the insect back to the next flower. Flowers of wind pollinated plants tend to lack petals and or sepals; typically large amounts of pollen are produced and pollination often occurs early in the growing season before leaves can interfere with the dispersal of the pollen. Many trees and all grasses and sedges are wind pollinated, as such they have no need for large fancy flowers.

An orchid flower

Plants have a number of different means to attract pollinators including colour, scent, heat, nectar glands, edible pollen and flower shape. Along with modifications involving the above structures two other conditions play a very important role in the sexual reproduction of flowering plants, the first is timing of flowering and the other is the size or number of flowers produced. Often plant species have a few large, very showy flowers while others produce many small flowers, often flowers are collected together into large inflorescences to maximize their visual effect, becoming more noticeable to passing pollinators. Flowers are attraction strategies and sexual expressions are functional strategies used to produce the next generation of plants, with pollinators and plants having co-evolved, often to some extraordinary degrees, very often rendering mutual benefit.

Flower heads showing disk and ray florets.

The largest family of flowering plants is the orchids (Orchidaceae), estimated by some specialists to include up to 35,000 species, which often have highly specialized flowers that attract particular insects for pollination. The stamens are modified to produce pollen in clusters called pollinia, which become attached to insects that crawl into the flower. The flower shapes may force insects to pass by the pollen, which is "glued" to the insect. Some orchids are even more highly specialized, with flower shapes that mimic the shape of insects to attract them to 'mate' with the flowers, a few even have scents that mimic insect pheromones.

Another large group of flowering plants is the Asteraceae or sunflower family with close to 22,000 species, which also have highly modified inflorescences that are flowers collected together in heads composed of a composite of individual flowers called florets. Heads with florets of one sex, when the flowers are pistillate or functionally staminate, or made up of all bisexual florets, are called homogamous and can include discoid and liguliflorous type heads. Some radiate heads may be homogamous too. Plants with heads that have florets of two or more sexual forms are called heterogamous and include radiate and disciform head forms, though some radiate heads may be heterogamous too.

Ferns

Ferns typically produce large diploid sporophytes with rhizomes, roots and leaves; and on fertile leaves called sporangium, spores are produced. The spores are released and germinate to produce short, thin gametophytes that are typically heart shaped, small and green in color. The gametophytes or thallus, produce both motile sperm in the antheridia and egg cells in separate archegonia. After rains or when dew deposits a film of water, the motile sperm are splashed away from the antheridia, which are normally produced on the top side of the thallus, and swim in the film of water to the antheridia where they fertilize the egg. To promote out crossing or cross fertilization the sperm are released before the eggs are receptive of the sperm, making it more likely that the sperm will fertilize the eggs of different thallus. A zygote is formed after fertilization, which grows into a new sporophytic plant. The condition of having separate sporophyte and gametophyte plants is call alternation of generations. Other plants with similar reproductive means include the *Psilotum*, *Lycopodium*, *Selaginella* and *Equisetum*.

Bryophytes

The bryophytes, which include liverworts, hornworts and mosses, reproduce both sexually and vegetatively. The gametophyte is the most commonly known phase of the plant. All are small plants found growing in moist locations and like ferns, have motile sperm with flagella and need water to facilitate sexual reproduction. These plants start as a haploid spore that grows into the dominate form, which is a multicellular haploid body with leaf-like structures that photosynthesize. Haploid gametes are produced in antherida and archegonia by mitosis. The sperm released from the antheridia respond to chemicals released by ripe archegonia and swim to them in a film of water and fertilize the egg cells, thus producing zygotes that are diploid. The zygote divides by mitotic division and grows into a sporophyte that is diploid. The multicellular diploid sporophyte produces structures called spore capsules. The spore capsules produce spores by meiosis, and when ripe, the capsules burst open and the spores are released. Bryophytes show considerable variation in their breeding structures and the above is a basic outline. In some species each gametophyte is one sex while other species produce both antheridia and archegonia on the same gametophyte which is thus hermaphrodite.

Sexual Morphology

Many plants have evolved complex sexual reproductive systems, which is expressed in different combinations of their reproductive organs. Some species have separate male and female plants, and some have separate male and female flowers on the same plant, but the majority of plants have both male and female parts in the same flower. Some plants change their morphological expression depending on a number of factors like age, time of day, or because of environmental conditions. Plant sexual morphology also varies within different populations of some species.

Certation

In botany, certation is competition in the style between pollen tubes attempting to fertilise the ovules there. If different pollen genotypes have different success rates then the genotype frequencies of the fertilised seeds will deviate from that which would occur if all pollen was equally successful. This process has been proposed to explain female-biased sex ratios in some dioecious plants, if female-determining pollen (which causes seeds to be female) is more successful than male-determining pollen (which causes seeds to be male), though other mechanisms whereby the sex ratio may be skewed away from 50:50 are also known.

Diaspore (Botany)

In botany, a diaspore is a plant dispersal unit consisting of a seed or spore *plus* any additional tissues that assist dispersal. In some seed plants, the diaspore is a seed and fruit together, or a seed and elaiosome. In a few seed plants, the diaspore is most or all of the plant, and is known as a tumbleweed.

Diaspores are common in weedy and ruderal species. Collectively, diaspores, seeds, and spores that have been modified for migration are disseminules.

Role in Dispersal

A diaspore of seed plus elaiosome is a common adaptation to seed dispersal by ants (myrmecochory). This is most notable in Australian and South African sclerophyll plant communities. Typically, ants carry the diaspore to their nest, where they may eat the elaiosome and discard the seed, and the seed may subsequently germinate.

Achenes of a dandelion (Taraxacum)

A diaspore of seed(s) plus fruit is common in plants dispersed by frugivores. Fruit-eating bats typically carry the diaspore to a favorite perch, where they eat the fruit and discard the seed. Fruit-eating birds typically swallow small seeds but, like bats, may carry larger seeded fruits to a perch where they eat the fruit and discard the seed. Diaspores such as achenes and samarae are dispersed primarily by wind; samaras are dispersed also by sailing or tumbling as they fall in still air. Drift fruits and some others are dispersed by water.

Tumbleweeds are dispersed by wind, sometimes over very long distances. These occur in a variety of weedy and ruderal species native to steppes and deserts. Grasses have various units of dispersal: rarely the caryopsis alone, often a diaspore. Disarticulation occurs below, between, or above the glumes and at all nodes. Although in some species the diaspore is a foxtail, in a few (the "tumble grasses") it is like a tumbleweed.

Flower

A poster with flowers or clusters of flowers produced by twelve species of flowering plants from different families

Various flowers from different families.

A flower, sometimes known as a bloom or blossom, is the reproductive structure found in plants that are floral (plants of the division Magnoliophyta, also called angiosperms). The biological function of a flower is to effect reproduction, usually by providing a mechanism for the union of sperm with eggs. Flowers may facilitate outcrossing (fusion of sperm and eggs from different individuals in a population) or allow selfing (fusion of sperm and egg from the same flower). Some flowers produce diaspores without fertilization (parthenocarpy). Flowers contain sporangia and are the site where gametophytes develop. Many flowers have evolved to be attractive to animals, so as to cause them to be vectors for the transfer of pollen. After fertilization, the ovary of the flower develops into fruit containing seeds.

In addition to facilitating the reproduction of flowering plants, flowers have long been admired and used by humans to beautify their environment, and also as objects of romance, ritual, religion, medicine and as a source of food.

Morphology

Main parts of a mature flower (Ranunculus glaberrimus)

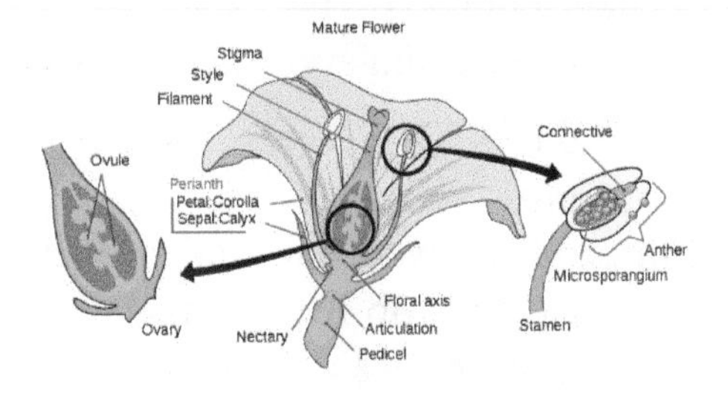

Diagram of flower parts

Floral parts

The essential parts of a flower can be considered in two parts: the vegetative part, consisting of petals and associated structures in the perianth, and the reproductive or sexual parts. A stereotypical flower consists of four kinds of structures attached to the tip of a short stalk. Each of these kinds of parts is arranged in a whorl on the receptacle. The four main whorls (starting from the base of the flower or lowest node and working upwards) are as follows:

Perianth

Collectively the calyx and corolla form the perianth.

- *Calyx*: the outermost whorl consisting of units called *sepals*; these are typically green and enclose the rest of the flower in the bud stage, however, they can be absent or prominent and petal-like in some species.

- *Corolla*: the next whorl toward the apex, composed of units called *petals*, which are typically thin, soft and colored to attract animals that help the process of pollination.

Reproductive

Reproductive parts of Easter Lily (*Lilium longiflorum*). 1. Stigma, 2. Style, 3. Stamens, 4. Filament, 5. Petal

- *Androecium*: the next whorl (sometimes multiplied into several whorls), consisting of units called stamens. Stamens consist of two parts: a stalk called a filament, topped by an anther where pollen is produced by meiosis and eventually dispersed.
- *Gynoecium*: the innermost whorl of a flower, consisting of one or more units called carpels. The carpel or multiple fused carpels form a hollow structure called an ovary, which produces ovules internally. Ovules are megasporangia and they in turn produce megaspores by meiosis which develop into female gametophytes. These give rise to egg cells. The gynoecium of a flower is also described using an alternative terminology wherein the structure one sees in the innermost whorl (consisting of an ovary, style and stigma) is called a pistil. A pistil may consist of a single carpel or a number of carpels fused together. The sticky tip of the pistil, the stigma, is the receptor of pollen. The supportive stalk, the style, becomes the pathway for pollen tubes to grow from pollen grains adhering to the stigma. The relationship to the gynoecium on the receptacle is described as hypogynous (beneath a superior ovary), perigynous (surrounding a superior ovary), or epigynous (above inferior ovary).

Structure

Although the arrangement described above is considered "typical", plant species show a wide variation in floral structure. These modifications have significance in the evolution of flowering plants and are used extensively by botanists to establish relationships among plant species.

The four main parts of a flower are generally defined by their positions on the receptacle and not by their function. Many flowers lack some parts or parts may be modified into other functions and/or look like what is typically another part. In some families, like Ranunculaceae, the petals are greatly reduced and in many species the sepals are colorful and petal-like. Other flowers have modified stamens that are petal-like; the double flowers of Peonies and Roses are mostly petaloid stamens. Flowers show great variation and plant scientists describe this variation in a systematic way to identify and distinguish species.

Specific terminology is used to describe flowers and their parts. Many flower parts are fused together; fused parts originating from the same whorl are connate, while fused parts originating from different whorls are adnate; parts that are not fused are free. When petals are fused into a tube or ring that falls away as a single unit, they are sympetalous (also called gamopetalous). Connate petals may have distinctive regions: the cylindrical base is the tube, the expanding region is the throat and the flaring outer region is the limb. A sympetalous flower, with bilateral symmetry with an upper and lower lip, is bilabiate. Flowers with connate petals or sepals may have various shaped corolla or calyx, including campanulate, funnelform, tubular, urceolate, salverform or rotate.

Referring to "fusion," as it is commonly done, appears questionable because at least some of the processes involved may be non-fusion processes. For example, the addition of intercalary growth at or below the base of the primordia of floral appendages such as sepals, petals, stamens and carpels may lead to a common base that is not the result of fusion.

Left: A normal zygomorphic Streptocarpus flower. Right: An aberrant peloric Streptocarpus flower. Both of these flowers appeared on the Streptocarpus hybrid 'Anderson's Crows' Wings'.

Many flowers have a symmetry. When the perianth is bisected through the central axis from any point, symmetrical halves are produced, forming a radial symmetry. These flowers are also known to be actinomorphic or regular, e.g. rose or trillium. When flowers are bisected and produce only one line that produces symmetrical halves the flower is said to be irregular or zygomorphic, e.g. snapdragon or most orchids.

Flowers may be directly attached to the plant at their base (sessile—the supporting stalk or stem is highly reduced or absent). The stem or stalk subtending a flower is called a peduncle. If a peduncle supports more than one flower, the stems connecting each flower to the main axis are called pedicels. The apex of a flowering stem forms a terminal swelling which is called the *torus* or receptacle.

Inflorescence

In those species that have more than one flower on an axis, the collective cluster of flowers is termed an *inflorescence*. Some inflorescences are composed of many small flowers arranged in a formation that resembles a single flower. The common example of this is most members of the very large composite (Asteraceae) group. A single daisy or sunflower, for example, is not a flower but a flower *head*—an inflorescence composed of numerous flowers (or florets). An inflorescence may include specialized stems and modified leaves known as bracts.

The familiar calla lily is not a single flower. It is actually an inflorescence of tiny flowers pressed together on a central stalk that is surrounded by a large petal-like bract.

Floral Diagrams and Floral Formulae

A *floral formula* is a way to represent the structure of a flower using specific letters, numbers and symbols, presenting substantial information about the flower in a compact form. It can represent a taxon, usually giving ranges of the numbers of different organs, or particular species. Floral formulae have been developed in the early 19th century and their use has declined since. Prenner *et al.* (2010) devised an extension of the existing model to broaden the descriptive capability of the formula. The format of floral formulae differs in different parts of the world, yet they convey the same information.

The structure of a flower can also be expressed by the means of *floral diagrams*. The use of schematic diagrams can replace long descriptions or complicated drawings as a tool for understanding both floral structure and evolution. Such diagrams may show important features of flowers, including the relative positions of the various organs, including the presence of fusion and symmetry, as well as structural details.

Development

A flower develops on a modified shoot or *axis* from a determinate apical meristem (*determinate* meaning the axis grows to a set size). It has compressed internodes, bearing structures that in classical plant morphology are interpreted as highly modified leaves. Detailed developmental studies, however, have shown that stamens are often initiated more or less like modified stems (caulomes) that in some cases may even resemble branchlets. Taking into account the whole diversity in the development of the androecium of flowering plants, we find a continuum between modified leaves (phyllomes), modified stems (caulomes), and modified branchlets (shoots).

Flowering Transition

The transition to flowering is one of the major phase changes that a plant makes during its life cycle. The transition must take place at a time that is favorable for fertilization and the formation of seeds, hence ensuring maximal reproductive success. To meet these needs a plant is able to interpret important endogenous and environmental cues such as changes in levels of plant hormones and seasonable temperature and photoperiod changes. Many perennial and most biennial plants require vernalization to flower. The molecular interpretation of these signals is through the transmission of a complex signal known as florigen, which involves a variety of genes, including CONSTANS, FLOWERING LOCUS C and FLOWERING LOCUS T. Florigen is produced in the leaves in reproductively favorable conditions and acts in buds and growing tips to induce a number of different physiological and morphological changes.

The first step of the transition is the transformation of the vegetative stem primordia into floral primordia. This occurs as biochemical changes take place to change cellular differentiation of leaf, bud and stem tissues into tissue that will grow into the reproductive organs. Growth of the central part of the stem tip stops or flattens out and the sides develop protuberances in a whorled or spiral fashion around the outside of the stem end. These protuberances develop into the sepals, petals, stamens, and carpels. Once this process begins, in most plants, it cannot be reversed and the stems develop flowers, even if the initial start of the flower formation event was dependent of some environmental cue. Once the process begins, even if that cue is removed the stem will continue to develop a flower.

Organ Development

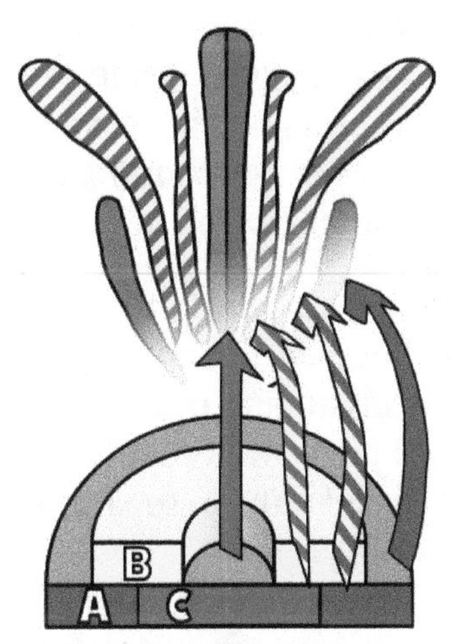

The ABC model of flower development

The molecular control of floral organ identity determination appears to be fairly well understood in some species. In a simple model, three gene activities interact in a combinatorial manner to determine the developmental identities of the organ primordia within the floral meristem. These gene functions are called A, B and C-gene functions. In the first floral whorl only A-genes are expressed, leading to the formation of sepals. In the second whorl both A- and B-genes are expressed, leading to the formation of petals. In the third whorl, B and C genes interact to form stamens and in the center of the flower C-genes alone give rise to carpels. The model is based upon studies of homeotic mutants in *Arabidopsis thaliana* and snapdragon, *Antirrhinum majus*. For example, when there is a loss of B-gene function, mutant flowers are produced with sepals in the first whorl as usual, but also in the second whorl instead of the normal petal formation. In the third whorl the lack of B function but presence of C-function mimics the fourth whorl, leading to the formation of carpels also in the third whorl.

Most genes central in this model belong to the MADS-box genes and are transcription factors that regulate the expression of the genes specific for each floral organ.

Floral Function

An example of a "perfect flower", this *Crateva religiosa* flower has both stamens (outer ring) and a pistil (center).

The principal purpose of a flower is the reproduction of the individual and the species. All flowering plants are *heterosporous*, producing two types of spores. Microspores are produced by meiosis inside anthers while megaspores are produced inside ovules, inside an ovary. In fact, anthers typically consist of four microsporangia and an ovule is an integumented megasporangium. Both types of spores develop into gametophytes inside sporangia. As with all heterosporous plants, the gametophytes also develop inside the spores (are endosporic).

In the majority of species, individual flowers have both functional carpels and stamens. Botanists describe these flowers as being *perfect* or *bisexual* and the species as *hermaphroditic*. Some flowers lack one or the other reproductive organ and called *imperfect* or *unisexual*. If unisex flowers are found on the same individual plant but in different locations, the species is said to be *monoecious*. If each type of unisex flower is found only on separate individuals, the plant is *dioecious*.

Flower Specialization and Pollination

Flowering plants usually face selective pressure to optimize the transfer of their pollen, and this is typically reflected in the morphology of the flowers and the behaviour of the plants. Pollen may be transferred between plants via a number of 'vectors'. Some plants make use of abiotic vectors — namely wind (anemophily) or, much less commonly, water (hydrophily). Others use biotic vectors including insects (entomophily), birds (ornithophily), bats (chiropterophily) or other animals. Some plants make use of multiple vectors, but many are highly specialised.

Cleistogamous flowers are self-pollinated, after which they may or may not open. Many Viola and some Salvia species are known to have these types of flowers.

The flowers of plants that make use of biotic pollen vectors commonly have glands called nectaries that act as an incentive for animals to visit the flower. Some flowers have patterns, called nectar guides, that show pollinators where to look for nectar. Flowers also attract pollinators by scent and color. Still other flowers use mimicry to attract pollinators. Some species of orchids, for example, produce flowers resembling female bees in color, shape, and scent. Flowers are also specialized in shape and have an arrangement of the stamens that ensures that pollen grains are transferred to the bodies of the pollinator when it lands in search of its attractant (such as nectar, pollen, or a mate). In pursuing this attractant from many flowers of the same species, the pollinator transfers pollen to the stigmas—arranged with equally pointed precision—of all of the flowers it visits.

Anemophilous flowers use the wind to move pollen from one flower to the next. Examples include grasses, birch trees, ragweed and maples. They have no need to attract pollinators and therefore tend not to be "showy" flowers. Male and female reproductive organs are generally found in separate flowers, the male flowers having a number of long filaments terminating in exposed stamens, and the female flowers having long, feather-like stigmas. Whereas the pollen of animal-pollinated flowers tends to be large-grained, sticky, and rich in protein (another "reward" for pollinators), anemophilous flower pollen is usually small-grained, very light, and of little nutritional value to animals.

Pollination

Grains of pollen sticking to this bee will be transferred to the next flower it visits

The primary purpose of a flower is reproduction. Since the flowers are the reproductive organs of plant, they mediate the joining of the sperm, contained within pollen, to the ovules — contained in the ovary. Pollination is the movement of pollen from the anthers to the stigma. The joining of the sperm to the ovules is called fertilization. Normally pollen is moved from one plant to another, but many plants are able to self pollinate. The fertilized ovules produce seeds that are the next generation. Sexual reproduction produces genetically unique offspring, allowing for adaptation. Flowers have specific designs which encourages the transfer of pollen from one plant to another of the same species. Many plants are dependent upon external factors for pollination, including: wind and animals, and especially insects. Even large animals such as birds, bats, and pygmy possums can be employed. The period of time during which this process can take place (the flower is fully expanded and functional) is called *anthesis*. The study of pollination by insects is called *anthecology*.

Pollen

The types of pollen that most commonly cause allergic reactions are produced by the plain-looking plants (trees, grasses, and weeds) that do not have showy flowers. These plants make small, light, dry pollen grains that are custom-made for wind transport.

The type of allergens in the pollen is the main factor that determines whether the pollen is likely to cause hay fever. For example, pine tree pollen is produced in large amounts by a common tree, which would make it a good candidate for causing allergy. It is, however, a relatively rare cause of allergy because the types of allergens in pine pollen appear to make it less allergenic.

Among North American plants, weeds are the most prolific producers of allergenic pollen. Ragweed is the major culprit, but other important sources are sagebrush, redroot pigweed, lamb's quarters, Russian thistle (tumbleweed), and English plantain.

There is much confusion about the role of flowers in allergies. For example, the showy and entomophilous goldenrod (*Solidago*) is frequently blamed for respiratory allergies, of which it is innocent, since its pollen cannot be airborne. Instead the allergen is usually the pollen of the contemporary bloom of anemophilous ragweed (*Ambrosia*), which can drift for many kilometers.

Scientists have collected samples of ragweed pollen 400 miles out at sea and 2 miles high in the air. A single ragweed plant can generate a million grains of pollen per day.

It is common to hear people say they are allergic to colorful or scented flowers like roses. In fact, only florists, gardeners, and others who have prolonged, close contact with flowers are likely to be sensitive to pollen from these plants. Most people have little contact with the large, heavy, waxy pollen grains of such flowering plants because this type of pollen is not carried by wind but by insects such as butterflies and bees.

Attraction Methods

A Bee orchid has evolved over many generations to better mimic a female bee to attract male bees as pollinators.

Plants cannot move from one location to another, thus many flowers have evolved to attract animals to transfer pollen between individuals in dispersed populations. Flowers that are insect-pollinated are called *entomophilous*; literally "insect-loving" in Greek. They can be highly modified along with the pollinating insects by co-evolution. Flowers commonly have glands called *nectaries* on various parts that attract animals looking for nutritious nectar. Birds and bees have color vision, enabling them to seek out "colorful" flowers.

Some flowers have patterns, called nectar guides, that show pollinators where to look for nectar; they may be visible only under ultraviolet light, which is visible to bees and some other insects.

Flowers also attract pollinators by scent and some of those scents are pleasant to our sense of smell. Not all flower scents are appealing to humans; a number of flowers are pollinated by insects that are attracted to rotten flesh and have flowers that smell like dead animals, often called Carrion flowers, including *Rafflesia*, the titan arum, and the North American pawpaw (*Asimina triloba*). Flowers pollinated by night visitors, including bats and moths, are likely to concentrate on scent to attract pollinators and most such flowers are white.

Other flowers use mimicry to attract pollinators. Some species of orchids, for example, produce flowers resembling female bees in color, shape, and scent. Male bees move from one such flower to another in search of a mate.

Pollination Mechanism

The pollination mechanism employed by a plant depends on what method of pollination is utilized.

Most flowers can be divided between two broad groups of pollination methods:

Entomophilous: flowers attract and use insects, bats, birds or other animals to transfer pollen from one flower to the next. Often they are specialized in shape and have an arrangement of the stamens that ensures that pollen grains are transferred to the bodies of the pollinator when it lands in search of its attractant (such as nectar, pollen, or a mate). In pursuing this attractant from many flowers of the same species, the pollinator transfers pollen to the stigmas—arranged with equally pointed precision—of all of the flowers it visits. Many flowers rely on simple proximity between flower parts to ensure pollination. Others, such as the *Sarracenia* or lady-slipper orchids, have elaborate designs to ensure pollination while preventing self-pollination.

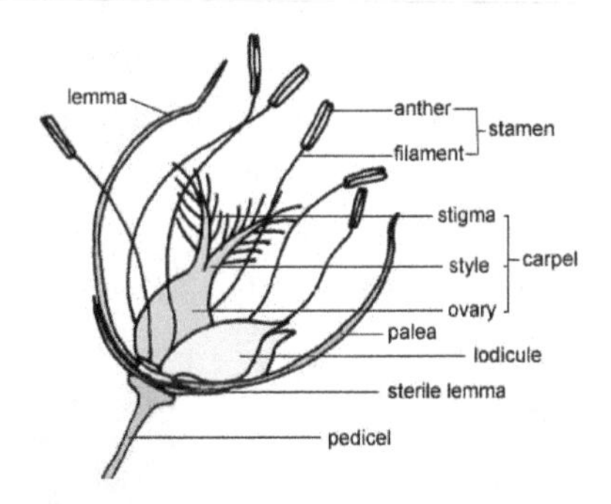

Grass flower with vestigial perianth or lodicules

Anemophilous: flowers use the wind to move pollen from one flower to the next, examples include the grasses, Birch trees, Ragweed and Maples. They have no need to attract pollinators and therefore tend not to grow large blossoms. Whereas the pollen of entomophilous flowers tends to be large-grained, sticky, and rich in protein (another "reward" for pollinators), anemophilous flower pollen is usually small-grained, very light, and of little nutritional value to insects, though it may still be gathered in times of dearth. Honeybees and bumblebees actively gather anemophilous corn (maize) pollen, though it is of little value to them.

Some flowers are self-pollinated and use flowers that never open or are self-pollinated before the flowers open, these flowers are called cleistogamous. Many Viola species and some Salvia have these types of flowers.

Flower-Pollinator Relationships

Many flowers have close relationships with one or a few specific pollinating organisms. Many flowers, for example, attract only one specific species of insect, and therefore rely on that insect for successful reproduction. This close relationship is often given as an example of coevolution, as the flower and pollinator are thought to have developed together over a long period of time to match each other's needs.

This close relationship compounds the negative effects of extinction. The extinction of either member in such a relationship would mean almost certain extinction of the other member as well. Some endangered plant species are so because of shrinking pollinator populations.

Fertilization and Dispersal

Some flowers with both stamens and a pistil are capable of self-fertilization, which does increase the chance of producing seeds but limits genetic variation. The extreme case of self-fertilization occurs in flowers that always self-fertilize, such as many dandelions. Conversely, many species of plants have ways of preventing self-fertilization. Unisexual male and female flowers on the same plant may not appear or mature at the same time, or pollen from the same plant may be incapable of fertilizing its ovules. The latter flower types, which have chemical barriers to their own pollen, are referred to as self-sterile or self-incompatible.

Evolution

While land plants have existed for about 425 million years, the first ones reproduced by a simple adaptation of their aquatic counterparts: spores. In the sea, plants—and some animals—can simply scatter out genetic clones of themselves to float away and grow elsewhere. This is how early plants reproduced. But plants soon evolved methods of protecting these copies to deal with drying out and other abuse which is even more likely on land than in the sea. The protection became the seed, though it had not yet evolved the flower. Early seed-bearing plants include the ginkgo and conifers. The earliest fossil of a flowering plant, *Archaefructus liaoningensis*, is dated about 125 million years old.

Archaefructus liaoningensis, one of the earliest known flowering plants

Several groups of extinct gymnosperms, particularly seed ferns, have been proposed as the ancestors of flowering plants but there is no continuous fossil evidence showing exactly how flowers evolved. The apparently sudden appearance of relatively modern flowers in the fossil record posed such a problem for the theory of evolution that it was called an "abominable mystery" by Charles

Darwin. Recently discovered angiosperm fossils such as *Archaefructus*, along with further discoveries of fossil gymnosperms, suggest how angiosperm characteristics may have been acquired in a series of steps.

Recent DNA analysis (molecular systematics) shows that *Amborella trichopoda*, found on the Pacific island of New Caledonia, is the sister group to the rest of the flowering plants, and morphological studies suggest that it has features which may have been characteristic of the earliest flowering plants.

The general assumption is that the function of flowers, from the start, was to involve animals in the reproduction process. Pollen can be scattered without bright colors and obvious shapes, which would therefore be a liability, using the plant's resources, unless they provide some other benefit. One proposed reason for the sudden, fully developed appearance of flowers is that they evolved in an isolated setting like an island, or chain of islands, where the plants bearing them were able to develop a highly specialized relationship with some specific animal (a wasp, for example), the way many island species develop today. This symbiotic relationship, with a hypothetical wasp bearing pollen from one plant to another much the way fig wasps do today, could have eventually resulted in both the plant(s) and their partners developing a high degree of specialization. Island genetics is believed to be a common source of speciation, especially when it comes to radical adaptations which seem to have required inferior transitional forms. Note that the wasp example is not incidental; bees, apparently evolved specifically for symbiotic plant relationships, are descended from wasps.

Likewise, most fruit used in plant reproduction comes from the enlargement of parts of the flower. This fruit is frequently a tool which depends upon animals wishing to eat it, and thus scattering the seeds it contains.

While many such symbiotic relationships remain too fragile to survive competition with mainland organisms, flowers proved to be an unusually effective means of production, spreading (whatever their actual origin) to become the dominant form of land plant life.

Amborella trichopoda, the sister group to the rest of the flowering plants

While there is only hard proof of such flowers existing about 130 million years ago, there is some circumstantial evidence that they did exist up to 250 million years ago. A chemical used by plants to defend their flowers, oleanane, has been detected in fossil plants that old, including gigantopterids, which evolved at that time and bear many of the traits of modern, flowering plants, though they are not known to be flowering plants themselves, because only their stems and prickles have been found preserved in detail; one of the earliest examples of petrification.

The similarity in leaf and stem structure can be very important, because flowers are genetically just an adaptation of normal leaf and stem components on plants, a combination of genes normally responsible for forming new shoots. The most primitive flowers are thought to have had a variable number of flower parts, often separate from (but in contact with) each other. The flowers would have tended to grow in a spiral pattern, to be bisexual (in plants, this means both male and female parts on the same flower), and to be dominated by the ovary (female part). As flowers grew more advanced, some variations developed parts fused together, with a much more specific number and design, and with either specific sexes per flower or plant, or at least "ovary inferior".

Flower evolution continues to the present day; modern flowers have been so profoundly influenced by humans that many of them cannot be pollinated in nature. Many modern, domesticated flowers used to be simple weeds, which only sprouted when the ground was disturbed. Some of them tended to grow with human crops, and the prettiest did not get plucked because of their beauty, developing a dependence upon and special adaptation to human affection.

Color

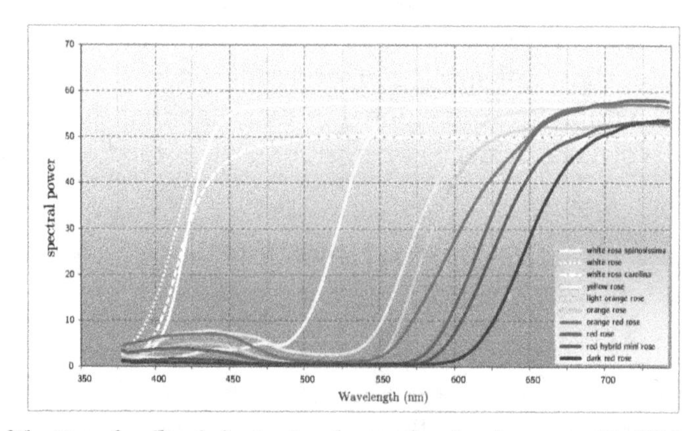

Spectrum of the flowers of the Rose family of plants. A red rose absorbs about 99.7% of light across a broad area below the red wavelengths of the spectrum, leading to an exceptionally *pure* red. A yellow rose will reflect about 5% of blue light, producing an unsaturated yellow (a yellow with a degree of white in it).

Many flowering plants reflect as much light as possible within the range of visible wavelengths of the pollinator the plant intends to attract. Flowers that reflect the full range of visible light are generally perceived as *white* by a human observer. An important feature of white flowers is that they reflect equally across the visible spectrum. While many flowering plants use white to attract pollinators, the use of color is also widespread (even within the same species). Color allows a flowering plant to be more specific about the pollinator it seeks to attract. The color model used by human color reproduction technology (CMYK) relies on the modulation of pigments that divide the spectrum into broad areas of absorption. Flowering plants by contrast are able to shift the transition point wavelength between absorption and reflection. If it is assumed that the visual systems of most pollinators view the visible spectrum as circular then it may be said that flowering plants produce color by absorbing the light in one region of the spectrum and reflecting the light in the other region. With CMYK, color is produced as a function of the amplitude of the broad regions of absorption. Flowering plants by contrast produce color by modifying the frequency (or rather wavelength) of the light reflected. Most flowers absorb light in the blue to yellow region of the spectrum and reflect light from the green to red region of the spectrum. For many species of flowering plant, it is the transition point that characterizes the color that they produce. Color may be modulated by shifting the transition point between absorption and reflection and in this way a flowering plant may specify which pollinator it seeks to attract.

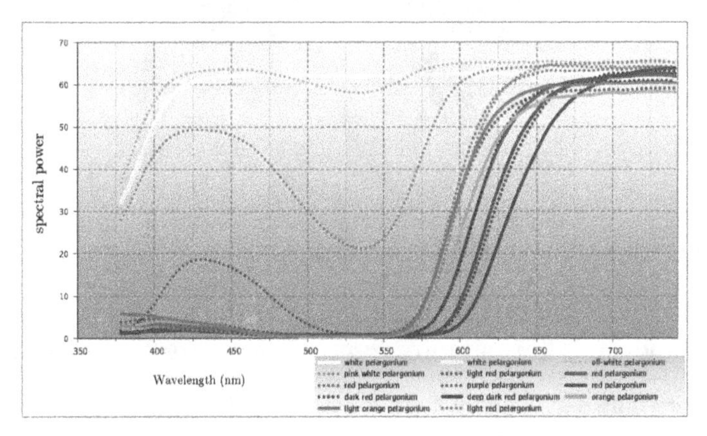

Spectrum of the flowers of the Pelargonium family of plants. Pelargonium produces exceptionally saturated colours in the orange to red range, with absorption of 99.6% across a broad area of the spectrum. Pelargonium is able to modulate its absorption producing pink with multiple levels of white.

Some flowering plants also have a limited ability to modulate areas of absorption. This is typically not as precise as control over wavelength. Humans observers will perceive this as degrees of saturation (the amount of *white* in the color).

Symbolism

Red Rose

Lilies are often used to denote life or resurrection

Flowers are common subjects of still life paintings, such as this one by Ambrosius Bosschaert the Elder

Many flowers have important symbolic meanings in Western culture. The practice of assigning meanings to flowers is known as floriography. Some of the more common examples include:

- Red roses are given as a symbol of love, beauty, and passion.
- Poppies are a symbol of consolation in time of death. In the United Kingdom, New Zealand, Australia and Canada, red poppies are worn to commemorate soldiers who have died in times of war.
- Irises/Lily are used in burials as a symbol referring to "resurrection/life". It is also associated with stars (sun) and its petals blooming/shining.
- Daisies are a symbol of innocence.

Flowers within art are also representative of the female genitalia, as seen in the works of artists such as Georgia O'Keeffe, Imogen Cunningham, Veronica Ruiz de Velasco, and Judy Chicago, and in fact in Asian and western classical art. Many cultures around the world have a marked tendency to associate flowers with femininity.

The great variety of delicate and beautiful flowers has inspired the works of numerous poets, especially from the 18th-19th century Romantic era. Famous examples include William Wordsworth's *I Wandered Lonely as a Cloud* and William Blake's *Ah! Sun-Flower*.

Because of their varied and colorful appearance, flowers have long been a favorite subject of visual artists as well. Some of the most celebrated paintings from well-known painters are of flowers, such as Van Gogh's sunflowers series or Monet's water lilies. Flowers are also dried, freeze dried and pressed in order to create permanent, three-dimensional pieces of flower art.

Their symbolism in dreams has also been discussed, with possible interpretations including "blossoming potential".

The Roman goddess of flowers, gardens, and the season of Spring is Flora. The Greek goddess of spring, flowers and nature is Chloris.

In Hindu mythology, flowers have a significant status. Vishnu, one of the three major gods in the Hindu system, is often depicted standing straight on a lotus flower. Apart from the association with Vishnu, the Hindu tradition also considers the lotus to have spiritual significance. For example, it figures in the Hindu stories of creation.

Usage

Flower market - Detroit's Eastern Market

A woman spreading flowers over a lingam in a temple in Varanasi

In modern times people have sought ways to cultivate, buy, wear, or otherwise be around flowers and blooming plants, partly because of their agreeable appearance and smell. Around the world, people use flowers for a wide range of events and functions that, cumulatively, encompass one's lifetime:

- For new births or christenings
- As a corsage or boutonniere worn at social functions or for holidays
- As tokens of love or esteem
- For wedding flowers for the bridal party, and for decorations for the hall
- As brightening decorations within the home
- As a gift of remembrance for *bon voyage* parties, welcome-home parties, and "thinking of you" gifts
- For funeral flowers and expressions of sympathy for the grieving

- For worshiping goddesses. In Hindu culture adherents commonly bring flowers as a gift to temples

People therefore grow flowers around their homes, dedicate entire parts of their living space to flower gardens, pick wildflowers, or buy flowers from florists who depend on an entire network of commercial growers and shippers to support their trade.

Flowers provide less food than other major plants parts (seeds, fruits, roots, stems and leaves) but they provide several important foods and spices. Flower vegetables include broccoli, cauliflower and artichoke. The most expensive spice, saffron, consists of dried stigmas of a crocus. Other flower spices are cloves and capers. Hops flowers are used to flavor beer. Marigold flowers are fed to chickens to give their egg yolks a golden yellow color, which consumers find more desirable; dried and ground marigold flowers are also used as a spice and colouring agent in Georgian cuisine. Flowers of the dandelion and elder are often made into wine. Bee pollen, pollen collected from bees, is considered a health food by some people. Honey consists of bee-processed flower nectar and is often named for the type of flower, e.g. orange blossom honey, clover honey and tupelo honey.

Hundreds of fresh flowers are edible but few are widely marketed as food. They are often used to add color and flavor to salads. Squash flowers are dipped in breadcrumbs and fried. Edible flowers include nasturtium, chrysanthemum, carnation, cattail, honeysuckle, chicory, cornflower, canna, and sunflower. Some edible flowers are sometimes candied such as daisy, rose, and violet (one may also come across a candied pansy).

Flowers can also be made into herbal teas. Dried flowers such as chrysanthemum, rose, jasmine, camomile are infused into tea both for their fragrance and medical properties. Sometimes, they are also mixed with tea leaves for the added fragrance.

Flowers have been used since as far back as 50,000 years in funeral rituals. Many cultures do draw a connection between flowers and life and death, and because of their seasonal return flowers also suggest rebirth, which may explain why many people place flowers upon graves. In ancient times the Greeks would place a crown of flowers on the head of the deceased as well as cover tombs with wreaths and flower petals. Rich and powerful women in ancient Egypt would wear floral head-dresses and necklaces upon their death as representations of renewal and a joyful afterlife, and the Mexicans to this day use flowers prominently in their Day of the Dead celebrations in the same way that their Aztec ancestors did.

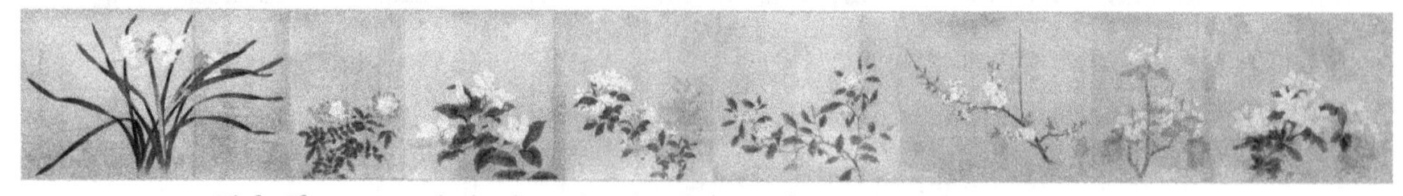

Eight Flowers, a painting by artist Qian Xuan, 13th century, Palace Museum, Beijing.

Gynoecium

Gynoecium (from Ancient Greek, *gyne*, meaning *woman*, and *oikos*, meaning *house*) is most commonly used as a collective term for the parts of a flower that produce ovules and ultimately develop into the fruit and seeds. The gynoecium is the innermost whorl of (one or more) pistils in a flower and is typically surrounded by the pollen-producing reproductive organs, the stamens, collectively called the androecium. The gynoecium is often referred to as the "female" portion of the flower, although rather than directly producing female gametes (i.e. egg cells), the gynoecium produces megaspores, each of which develops into a female gametophyte which then produces egg cells.

Flower of *Magnolia × wieseneri* showing the many pistils making up the gynoecium in the middle of the flower

Hippeastrum flowers showing stamens, style and stigma

Hippeastrum stigmas and style

The term gynoecium is also used by botanists to refer to a cluster of archegonia and any associated modified leaves or stems present on a gametophyte shoot in mosses, liverworts and hornworts. The corresponding terms for the male parts of those plants are clusters of antheridia within the androecium.

Moss plants with gynoecia, clusters of archegonia at the apex of each shoot.

Flowers that bear a gynoecium but no stamens are called carpellate. Flowers lacking a gynoecium are called staminate.

The gynoecium is often referred to as female because it gives rise to female (egg-producing) gametophytes, however, strictly speaking sporophytes do not have sex, only gametophytes do.

Gynoecium development and arrangement is important in systematic research and identification of angiosperms, but can be the most challenging of the floral parts to interpret.

Pistils

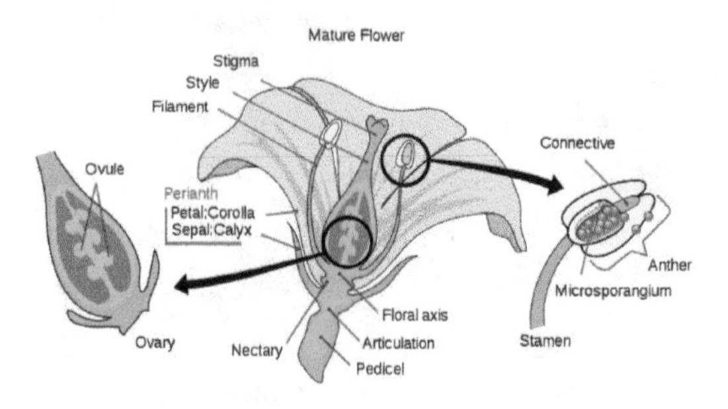

A monocarpous (single carpel) gynoecium in context. The gynoecium (whether composed of a single carpel or multiple "fused" carpels) is typically made up of an ovary, style, and stigma as in the center of the flower.

The gynoecium may consist of one or more separate pistils. A pistil typically consists of an expanded basal portion called the ovary, an elongated section called a style and an apical structure that receives pollen called a stigma.

- The ovary (from Latin *ovum* meaning egg), is the enlarged basal portion which contains placentas, ridges of tissue bearing one or more ovules (integumented megasporangia). The placentas and/or ovule(s) may be born on the gynoecial appendages or less frequently on the floral apex. The chamber in which the ovules develop is called a locule (or sometimes cell).
- The style (from Ancient Greek stülos meaning a pillar), is a pillar-like stalk through which pollen tubes grow to reach the ovary. Some flowers such as *Tulipa* do not have a distinct style, and the stigma sits directly on the ovary. The style is a hollow tube in some plants such as lilies, or has transmitting tissue through which the pollen tubes grow.

- The stigma (from Ancient Greek στίγμα, *stigma* meaning mark, or puncture), is usually found at the tip of the style, the portion of the carpel(s) that receives pollen (male gametophytes). It is commonly sticky or feathery to capture pollen.

The word "pistil" comes from Latin *pistillum* meaning pestle. A sterile pistil in a male flower is referred to as a pistillode.

Carpels

The pistils of a flower are considered to be composed of carpels. A carpel is a theoretical construct interpreted as modified leaves bearing structures called ovules, inside which the egg cells ultimately form. A pistil may consist of one carpel, with its ovary, style and stigma, or several carpels may be joined together with a single ovary, the whole unit called a pistil. The gynoecium may consist of one or more uni-carpellate (with one carpel) pistils, or of one multi-carpellate pistil.

Carpels are thought to be phylogenetically derived from ovule-bearing leaves or leaf homologues (megasporophylls), which evolved to form a closed structure containing the ovules. This structure is typically rolled and fused along the margin.

Although many flowers satisfy the above definition of a carpel, there are also flowers that do not have carpels according to this definition because in these flowers the ovule(s), although enclosed, are borne directly on the shoot apex, and only later become enclosed by the carpel. Different remedies have been suggested for this problem. An easy remedy that applies to most cases is to redefine the carpel as an appendage that encloses ovule(s) and may or may not bear them.

Centre of a *Ranunculus repens* (Creeping Buttercup) showing multiple unfused carpels surrounded by longer stamens

Cross-section through the ovary of *Narcissus* showing multiple connate carpels (a compound pistil) fused along the placental line where the ovules form in each locule

Types of Gynoecia

If a gynoecium has a single carpel, it is called monocarpous. If a gynoecium has multiple, distinct (free, unfused) carpels, it is apocarpous. If a gynoecium has multiple carpels "fused" into a single structure, it is syncarpous. A syncarpous gynoecium can sometimes appear very much like a monocarpous gynoecium.

Comparison of gynoecium terminology using carpel and pistil			
Gynoecium composition	Carpel terminology	Pistil terminology	Examples
Single carpel	Monocarpous (unicarpellate) gynoecium	A pistil (simple)	Avocado (Persea sp.), most legumes (Fabaceae)
Multiple distinct (unfused) carpels	Apocarpous (choricarpous) gynoecium	Pistils (simple)	Strawberry (Fragaria sp.), Buttercup (Ranunculus sp.)
Multiple connate ("fused") carpels	Syncarpous gynoecium	A pistil (compound)	Tulip (Tulipa sp.), most flowers

The degree of connation ("fusion") in a syncarpous gynoecium can vary. The carpels may be "fused" only at their bases, but retain separate styles and stigmas. The carpels may be "fused" entirely, except for retaining separate stigmas. Sometimes (e.g., Apocynaceae) carpels are fused by their styles or stigmas but possess distinct ovaries. In a syncarpous gynoecium, the "fused" ovaries of the constituent carpels may be referred to collectively as a single compound ovary. It can be a challenge to determine how many carpels fused to form a syncarpous gynoecium. If the styles and stigmas are distinct, they can usually be counted to determine the number of carpels. Within the compound ovary, the carpels may have distinct locules divided by walls called septa. If a syncarpous gynoecium has a single style and stigma and a single locule in the ovary, it may be necessary to examine how the ovules are attached. Each carpel will usually have a distinct line of placentation where the ovules are attached.

Pistil Development

Pistils begin as small primordia on a floral apical meristem, forming later than, and closer to the (floral) apex than sepal, petal and stamen primordia. Morphological and molecular studies of pistil ontogeny reveal that carpels are most likely homologous to leaves.

A carpel has a similar function to a megasporophyll, but typically includes a stigma, and is fused, with ovules enclosed in the enlarged lower portion, the ovary.

In some basal angiosperm lineages, Degeneriaceae and Winteraceae, a carpel begins as a shallow cup where the ovules develop with laminar placentation, on the upper surface of the carpel. The carpel eventually forms a folded, leaf-like structure, not fully sealed at its margins. No style exists, but a broad stigmatic crest along the margin allows pollen tubes access along the surface and between hairs at the margins.

Two kinds of fusion have been distinguished: postgenital fusion that can be observed during the development of flowers, and congenital fusion that cannot be observed i.e., fusions that occurred during phylogeny. But it is very difficult to distinguish fusion and non-fusion processes in the evolution of flowering plants. Some processes that have been considered congenital (phylogenetic) fusions appear to be non-fusion processes such as, for example, the de novo formation of intercalary growth in a ring zone at or below the base of primordia. Therefore, "it is now increasingly acknowledged that the term 'fusion,' as applied to phylogeny (as in 'congenital fusion') is ill-advised."

Gynoecium Position

Flowers and fruit (capsules) of the ground orchid, *Spathoglottis plicata*, illustrating an **inferior** ovary.

Basal angiosperm groups tend to have carpels arranged spirally around a conical or dome-shaped receptacle. In later lineages, carpels tend to be in whorls.

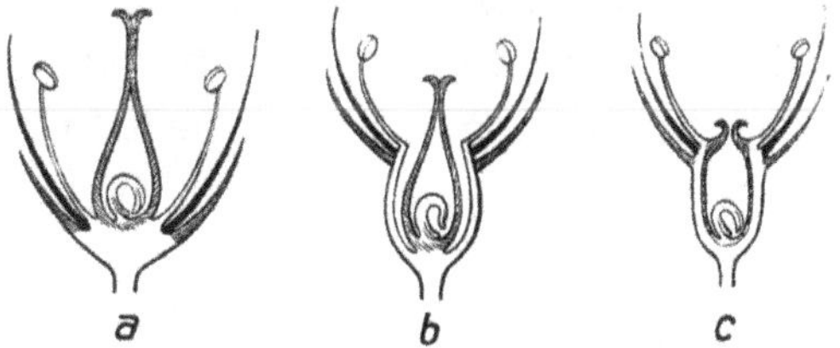

Illustration showing longitudinal sections through hypogynous (a), perigynous (b), and epigynous (c) flowers

The relationship of the other flower parts to the gynoecium can be an important systematic and taxonomic character. In some flowers, the stamens, petals, and sepals are often said to be "fused" into a "floral tube" or hypanthium. However, as Leins & Erbar (2010) pointed out, "the classical view that the wall of the inferior ovary results from the "congenital" fusion of dorsal carpel flanks and the floral axis does not correspond to the ontogenetic processes that can actually be observed. All that can be seen is an intercalary growth in a broad circular zone that changes the shape of the floral axis (receptacle)." And what happened during evolution is not a phylogenetic fusion but the formation of a unitary intercalary meristem. Evolutionary developmental biology investigates such developmental processes that arise or change during evolution.

If the hypanthium is absent, the flower is *hypogynous*, and the stamens, petals, and sepals are all attached to the receptacle below the gynoecium. Hypogynous flowers are often referred to as having a *superior ovary*. This is the typical arrangement in most flowers.

If the hypanthium is present up to the base of the style(s), the flower is *epigynous*. In an epigynous flower, the stamens, petals, and sepals are attached to the hypanthium at the top of the ovary or, occasionally, the hypanthium may extend beyond the top of the ovary. Epigynous flowers are often referred to as having an *inferior ovary*. Plant families with epigynous flowers include orchids, asters, and evening primroses.

Between these two extremes are *perigynous* flowers, in which a hypanthium is present, but is either free from the gynoecium (in which case it may appear to be a cup or tube surrounding the gynoecium) or connected partly to the gynoecium (with the stamens, petals, and sepals attached to the hypanthium part of the way up the ovary). Perigynous flowers are often referred to as having a *half-inferior ovary* (or, sometimes, *partially inferior* or *half-superior*). This arrangement is particularly frequent in the rose family and saxifrages.

Occasionally, the gynoecium is born on a stalk, called the gynophore, as in *Isomeris arborea*.

Placentation

Within the ovary, each ovule is born by a placenta or arises as a continuation of the floral apex. The placentas often occur in distinct lines called lines of placentation. In monocarpous or apocarpous gynoecia, there is typically a single line of placentation in each ovary. In syncarpous gynoecia, the lines of placentation can be regularly spaced along the wall of the ovary (parietal placentation), or near the center of the ovary. In the latter case, separate terms are used depending on whether or not the ovary is divided into separate locules. If the ovary is divided, with the ovules born on a line of placentation at the inner angle of each locule, this is axile placentation. An ovary with free central placentation, on the other hand, consists of a single compartment without septae and the ovules are attached to a central column that arises directly from the floral apex (axis). In some cases a single ovule is attached to the bottom or top of the locule (basal or apical placentation, respectively).

The Ovule

Longitudinal section of carpellate flower of squash showing ovary, ovules, stigma, style, and petals

In flowering plants, the *ovule* (from Latin *ovulum* meaning small egg) is a complex structure born inside ovaries. The ovule initially consists of a stalked, integumented *megasporangium* (also called the *nucellus*). Typically, one cell in the megasporangium undergoes meiosis resulting in one to four megaspores. These develop into a megagametophyte (often called the embryo sac) within the ovule. The megagametophyte typically develops a small number of cells, including two special cells, an egg cell and a binucleate central cell, which are the gametes involved in double fertilization. The central cell, once fertilized by a sperm cell from the pollen becomes the first cell of the endosperm, and the egg cell once fertilized become the zygote that develops into the embryo. The gap in the integuments through which the pollen tube enters to deliver sperm to the egg is called the micropyle. The stalk attaching the ovule to the placenta is called the funiculus.

Role of the Stigma and Style

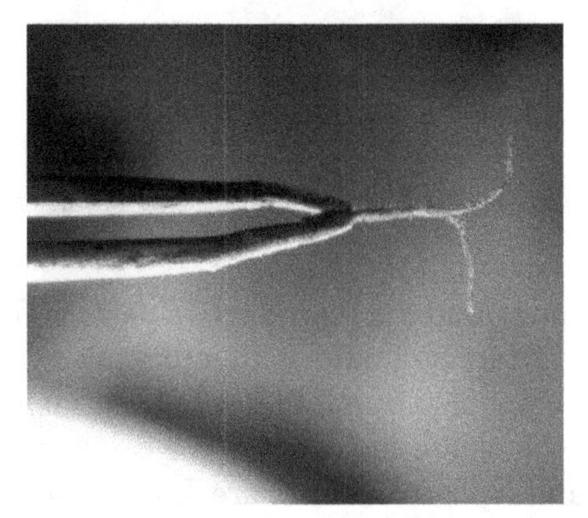

Stigmas and style of *Cannabis sativa* held in a pair of forceps

Stigma of a *Crocus* flower.

Stigmas can vary from long and slender to globe-shaped to feathery. The stigma is the receptive tip of the carpel(s), which receives pollen at pollination and on which the pollen grain germinates. The stigma is adapted to catch and trap pollen, either by combining pollen of visiting insects or by various hairs, flaps, or sculpturings.

The style and stigma of the flower are involved in most types of self incompatibility reactions. Self-incompatibility, if present, prevents fertilization by pollen from the same plant or from genetically similar plants, and ensures outcrossing.

Orbicule

Orbicules (syn. Ubisch bodies, con-peito grains) are small acellular structures of sporopollenin (known size range from < 1 µm to 15 µm, but usually sub-micrometre) that might occur on the inner tangential and radial walls of tapetal cells. Their function is unclear at this moment. Current consensus is that they are just a by-product of pollen wall sporopollenin synthesis.

Petal

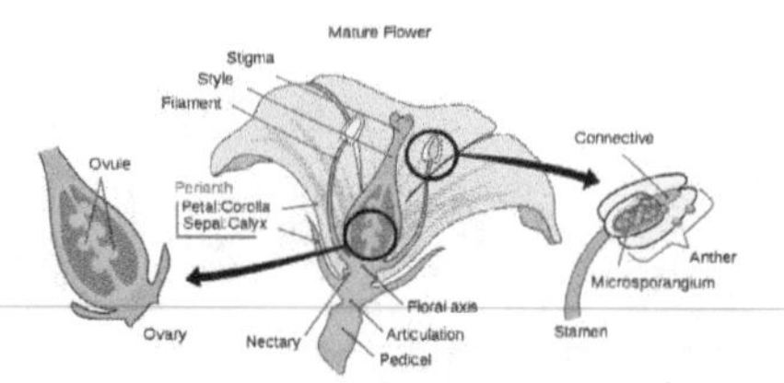

Diagram showing the parts of a mature flower. In this example the perianth is separated into a calyx (sepals) and corolla (petals)

Petals are modified leaves that surround the reproductive parts of flowers. They are often brightly colored or unusually shaped to attract pollinators. Together, all of the petals of a flower are called a *corolla*. Petals are usually accompanied by another set of special leaves called sepals, that collectively form the *calyx* and lie just beneath the corolla. The calyx and the corolla together make up the perianth. When the petals and sepals of a flower are difficult to distinguish, they are collectively called tepals. Examples of plants in which the term *tepal* is appropriate include genera such as *Aloe* and *Tulipa*. Conversely, genera such as *Rosa* and *Phaseolus* have well-distinguished sepals and petals. When the undifferentiated tepals resemble petals, they are referred to as "petaloid", as in petaloid monocots, orders of monocots with brightly coloured tepals. Since they include Liliales, an alternative name is lilioid monocots.

Although petals are usually the most conspicuous parts of animal-pollinated flowers, wind-pollinated species, such as the grasses, either have very small petals or lack them entirely.

Tetrameric flower of a Primrose willowherb (*Ludwigia octovalvis*) showing petals and sepals

A Tulip's actinomorphic flower with three petals and three sepals, that collectively present a good example of an undifferentiated perianth. In this case, the word "tepals" is used.

Corolla

Corolla forming a tube with long points (and a separate green calyx tube)

The role of the corolla in plant evolution has been studied extensively since Charles Darwin postulated a theory of the origin of elongated corollae and corolla tubes.

If the petals are free from one another in the corolla, the plant is *polypetalous* or *choripetalous*; while if the petals are at least partially fused together, it is *gamopetalous* or *sympetalous*. The corolla in some plants forms a tube.

Variations

Pelargonium peltatum: its floral structure is almost identical to that of geraniums, but it is conspicuously zygomorphic

Geranium incanum, with an actinomorphic flower typical of the genus

Petals can differ dramatically in different species. The number of petals in a flower may hold clues to a plant's classification. For example, flowers on eudicots (the largest group of dicots) most frequently have four or five petals while flowers on monocots have three or six petals, although there are many exceptions to this rule.

White pea, a zygomorphic flower

Narcissus pseudonarcissus, showing from the bend to the tip of the flower: spathe, floral cup, tepals, corona

The petal whorl or corolla may be either radially or bilaterally symmetrical. If all of the petals are essentially identical in size and shape, the flower is said to be regular or actinomorphic (meaning "ray-formed"). Many flowers are symmetrical in only one plane (i.e., symmetry is bilateral) and are termed irregular or zygomorphic (meaning "yoke-" or "pair-formed"). In *irregular* flowers, other floral parts may be modified from the *regular* form, but the petals show the greatest deviation from radial symmetry. Examples of zygomorphic flowers may be seen in orchids and members of the pea family.

In many plants of the aster family such as the sunflower, *Helianthus annuus*, the circumference of the flower head is composed of ray florets. Each ray floret is anatomically an individual flower with a single large petal. Florets in the centre of the disc typically have no or very reduced petals. In some plants such as *Narcissus* the lower part of the petals or tepals are fused to form a floral cup (hypanthium) above the ovary, and from which the petals proper extend.

Petal often consists of two parts: the upper, broad part, similar to leaf blade, called the *limb* and the lower part, narrow, similar to leaf petiole, called the *claw*. Claws are developed in petals of some flowers of the family *Brassicaceae*, such as *Erysimum cheiri*.

The inception and further development of petals shows a great variety of patterns. Petals of different species of plants vary greatly in colour or colour pattern, both in visible light and in ultraviolet. Such patterns often function as guides to pollinators, and are variously known as nectar guides, pollen guides, and floral guides.

Genetics

The genetics behind the formation of petals, in accordance with the ABC model of flower development, are that sepals, petals, stamens, and carpels are modified versions of each other. It appears that the mechanisms to form petals evolved very few times (perhaps only once), rather than evolving repeatedly from stamens.

Significance of Pollination

Pollination is an important step in the sexual reproduction of higher plants. Pollen is produced by the male flower or by the male organs of hermaphroditic flowers.

Pollen does not move on its own and thus requires wind or animal pollinators to disperse the pollen to the stigma (botany) of the same or nearby flowers. However, pollinators are rather selective in determining the flowers they choose to pollinate. This develops competition between flowers and as a result flowers must provide incentives to appeal to pollinators (unless the flower self-pollinates or is involved in wind pollination). Petals play a major role in competing to attract pollinators. Henceforth pollination dispersal could occur and the survival of many species of flowers could prolong.

Functions and Purposes of Petals

Petals have various functions and purposes depending on the type of plant. In general, petals operate to protect some parts of the flower and attract/repel specific pollinators. Flower Petal Function: This is where the positioning of the flower petals are located on the flower is the corolla e.g. the buttercup having shiny yellow flower petals which contain guidelines amongst the petals in aiding the pollinator towards the nectar. Pollinators have the ability to determine specific flowers they wish to pollinate. Using incentives flowers draw pollinators and set up a mutual relation between each other in which case the pollinators will remember to always guard and pollinate these flowers (unless incentives are not consistently met and competition prevails).

Scent

The petals could produce different scents to allure desirable pollinators and/or repel undesirable pollinators. Some flowers will also mimic the scents produced by materials such as decaying meat, to attract pollinators to them.

Colour

Various colour traits are used by different petals that could attract pollinators that have poor smelling abilities, or that only come out at certain parts of the day. Some flowers are able to change the colour of their petals as a signal to mutual pollinators to approach or keep away.

Shape and Size

Furthermore, the shape and size of the flower/petals is important in selecting the type of pollinators they need. For example, large petals and flowers will attract pollinators at a large distance and/or that are large themselves. Collectively the scent, colour and shape of petals all play a role in attracting/repelling specific pollinators and providing suitable conditions for pollinating. Some pollinators include insects, birds, bats and the wind. In some petals, a distinction can be made between a lower narrowed, stalk-like basal part referred to as the claw, and a wider distal part referred to as the blade. Often the claw and blade are at an angle with one another.

Types of Pollination

Wind Pollination

Wind-pollinated flowers often have small dull petals and produce little or no scent. Some of these flowers will often have no petals at all. Flowers that depend on wind pollination will produce large amounts of pollen because most of the pollen scattered by the wind tends to not reach other flowers.

Attracting Insects

Flowers have various regulatory mechanisms in order to attract insects. One such helpful mechanism is the use of colour guiding marks. Insects such as the bee or butterfly can see the ultraviolet marks which are contained on these flowers, acting as an attractive mechanism which is not visible towards the human eye. Many flowers contain a variety of shapes acting to aid with the landing of the visiting insect and also influence the insect to brush against anthers and stigmas (parts of the flower). One such example of a flower is the pōhutukawa (*Metrosideros excelsa*) which acts to regulate colour within a different way. The pōhutukawa contains small petals also having bright large red clusters of stamens. Another attractive mechanism for flowers is the use of scent which is highly attractive towards humans such as the rose, but some are very fragrant within attracting flies as they have a smell of rotting meat. Dark is another factor in which flowers have grown to adapt these conditions so colour lacks vision at night therefore scent is the solution for flowers which are pollinated by night flying insects such as the moth.

Attracting Birds

Flowers are also pollinated by birds and must be large and colorful to be visible against natural scenery. Such bird –pollinated native plants include: Kōwhai (*Sophora* species), flax (*Phormium tenax*, harakeke) and kākā beak (*Clianthus puniceus*, kōwhai ngutu-kākā). Interestingly enough, flowers adapt the mechanism on their petals to change colour in acting as a communicative mechanism for the bird to visit. An example is the tree fuchsia (*Fuchsia excorticata*, kōtukutuku) which are green when needing to be pollinated and turn red for the birds to stop coming and pollinating the flower.

Bat-Pollinated Flowers

Flowers can be pollinated by short tailed bats. An example of this is the dactylanthus (*Dactylanthus taylorii*). This plant has its home under the ground acting the role of a parasite on the roots of forest trees. The dactylanthus has only its flowers pointing to the surface and the flowers lack colour but have the advantage of containing lots of nectar and a very strong scent. These act as a very useful mechanism in attracting the bat.

Stamen

Stamens of a *Hippeastrum* with white filaments and prominent anthers carrying pollen

The stamen (plural *stamina* or *stamens*) is the pollen-producing reproductive organ of a flower. Collectively the stamens form the androecium.

Morphology and Terminology

A stamen typically consists of a stalk called the filament and an anther which contains *microsporangia*. Most commonly anthers are two-lobed and are attached to the filament either at the base or in the middle area of the anther. The sterile tissue between the lobes is called the connective. A pollen grain develops from a microspore in the microsporangium and contains the male gametophyte.

The stamens in a flower are collectively called the *androecium*. The androecium in various species of plants forms a great variety of patterns, some of them highly complex. It surrounds the gynoecium and is surrounded by the perianth. A few members of the family Triuridaceae, particularly *Lacandonia schismatica*, are exceptional in that their gynoecia surround their androecia.

Hippeastrum flowers showing stamens above the style with its terminal stigma

Etymology

- *Stamen* is the Latin word meaning "thread" (originally thread of the warp, in weaving).
- Filament derives from classical Latin *filum*, meaning "thread"
- Anther derives from French *anthère*, from classical Latin *anthera*, meaning "medicine extracted from the flower" in turn from Ancient "flower"

Variation in Morphology

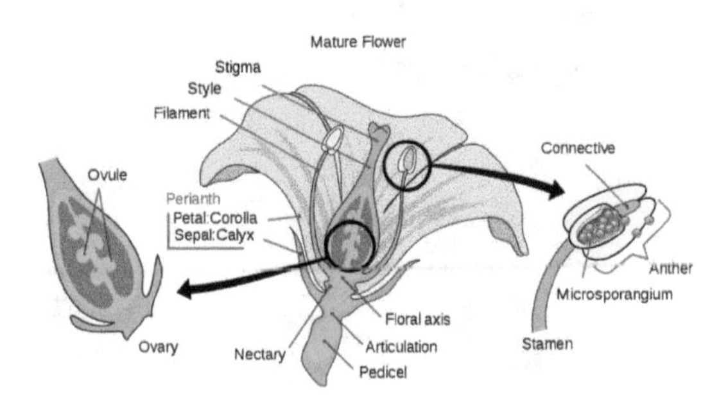

Stamens in context

Depending on the species of plant, some or all of the stamens in a flower may be attached to the petals or to the floral axis. They also may be free-standing or fused to one another in many different ways, including fusion of some but not all stamens. The filaments may be fused and the anthers free, or the filaments free and the anthers fused. Rather than there being two locules, one locule of a stamen may fail to develop, or alternatively the two locules may merge late in development to give a single locule. Extreme cases of stamen fusion occur in some species of *Cyclanthera* in the family Cucurbitaceae and in section *Cyclanthera* of genus *Phyllanthus* (family Euphorbiaceae) where the stamens form a ring around the gynoecium, with a single locule.

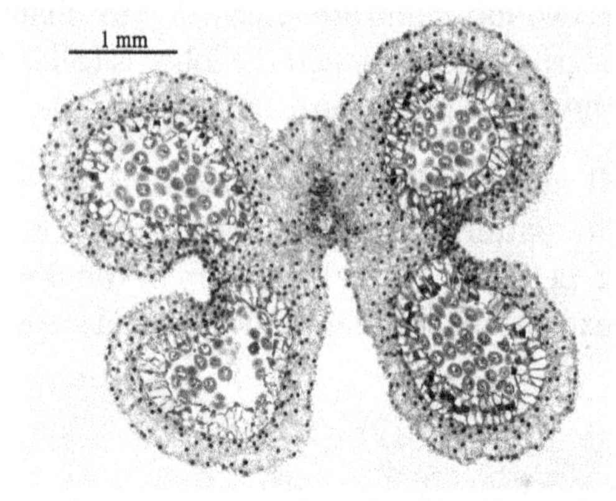

Cross section of a *Lilium* stamen, with four locules surrounded by the tapetum

Pollen Production

A typical anther contains four microsporangia. The *microsporangia* form sacs or pockets (*locules*) in the anther. The two separate locules on each side of an anther may fuse into a single locule. Each microsporangium is lined with a nutritive tissue layer called the *tapetum* and initially contains diploid pollen mother cells. These undergo meiosis to form haploid spores. The spores may remain attached to each other in a tetrad or separate after meiosis. Each microspore then divides mitotically to form an immature microgametophyte called a pollen grain.

The pollen is eventually released when the anther forms openings (dehisces). These may consist of longitudinal slits, pores, as in the heath family (Ericaceae), or by valves, as in the barberry family (Berberidaceae). In some plants, notably members of Orchidaceae and Asclepiadoideae, the pollen remains in masses called pollinia, which are adapted to attach to particular pollinating agents such as birds or insects. More commonly, mature pollen grains separate and are dispensed by wind or water, pollinating insects, birds or other pollination vectors.

Pollen of angiosperms must be transported to the stigma, the receptive surface of the *carpel*, of a compatible flower, for successful pollination to occur. After arriving, the pollen grain (an immature microgametophyte) typically completes its development. It may grow a pollen tube and undergoing mitosis to produce two sperm nuclei.

Sexual Reproduction in Plants

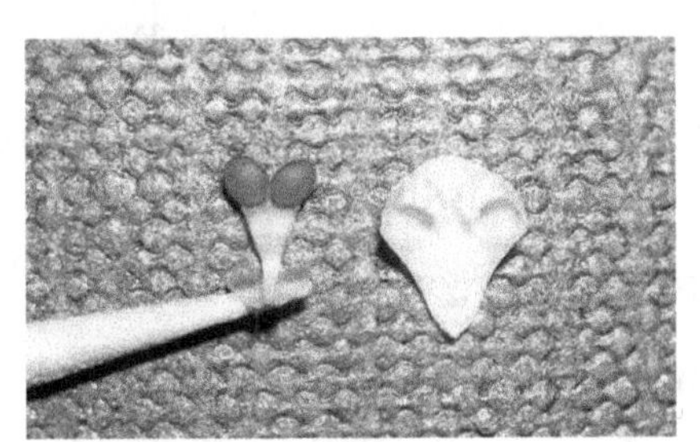

Stamen with pollinia and its anther cap. *Phalaenopsis* orchid.

In the typical flower (that is, in the majority of flowering plant species) each flower has both carpels and stamens. In some species, however, the flowers are unisexual with only carpels or stamens. (monoecious = both types of flowers found on the same plant; dioecious = the two types of flower found only on different plants). A flower with only stamens is called androecious. A flower with only carpels is called gynoecious.

A flower having only functional stamens and lacking functional carpels is called a staminate flower, or (inaccurately) male. A plant with only functional carpels is called pistillate, or (inaccurately) female.

An abortive or rudimentary stamen is called a staminodium or staminode, such as in *Scrophularia nodosa*.

The carpels and stamens of orchids are fused into a column. The top part of the column is formed by the anther, which is covered by an anther cap.

Descriptive Terms

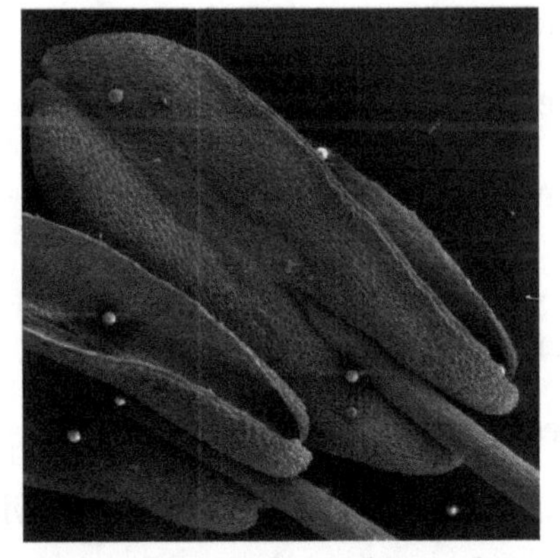

Scanning electron microscope image of *Pentas lanceolata* anthers, with pollen grains on surface

- A column formed from the fusion of multiple filaments is known as an *androphore*.

The anther can be attached to the filament's connective in two ways:

- basifixed: attached at its base to the filament
    - pseudobasifixed: a somewhat misnomer configuration where connective tissue extends in a tube around the filament
- dorsifixed: attached at its center to the filament, usually versatile (able to move)

Stamens can be connate (fused or joined in the same whorl):

- extrorse: anther dehiscence directed away from the centre of the flower. Cf. introrse, directed inwards, and latrorse towards the side.
- monadelphous: fused into a single, compound structure
- declinate: curving downwards, then up at the tip (also - declinate-descending)
- diadelphous: joined partially into two androecial structures
- pentadelphous: joined partially into five androecial structures
- synandrous: only the anthers are connate (such as in the Asteraceae). The fused stamens are referred to as a synandrium.

Stamens can also be adnate (fused or joined from more than one whorl):

- epipetalous: adnate to the corolla
- epiphyllous: adnate to undifferentiated tepals (as in many Liliaceae)

They can have different lengths from each other:

- didymous: two equal pairs
- didynamous: occurring in two pairs, a long pair and a shorter pair
- tetradynamous: occurring as a set of six stamens with four long and two shorter ones

or respective to the rest of the flower (perianth):

- exserted: extending beyond the corolla
- included: not extending beyond the corolla

They may be arranged in one of two different patterns:

- spiral; or
- whorled: one or more discrete whorls (series)

They may be arranged, with respect to the petals:

- diplostemonous: in two whorls, the outer alternating with the petals, while the inner is opposite the petals.

- obdiplostemonous: in two whorls, the outer opposite the petals

Stigma (Botany)

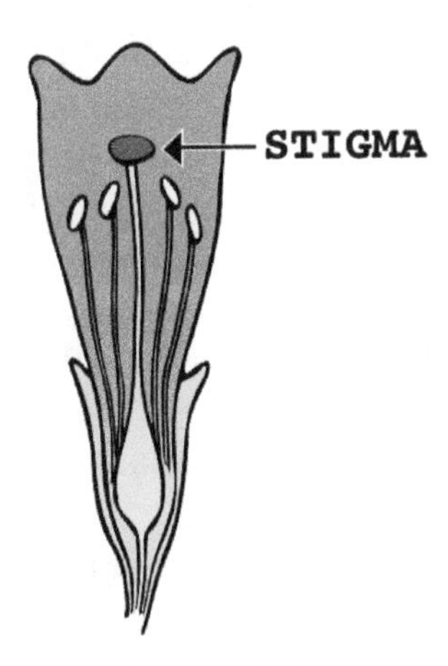

Diagram of stigma

The stigma (plural: stigmata) is the receptive tip of a carpel, or of several fused carpels, in the gynoecium of a flower.

Description

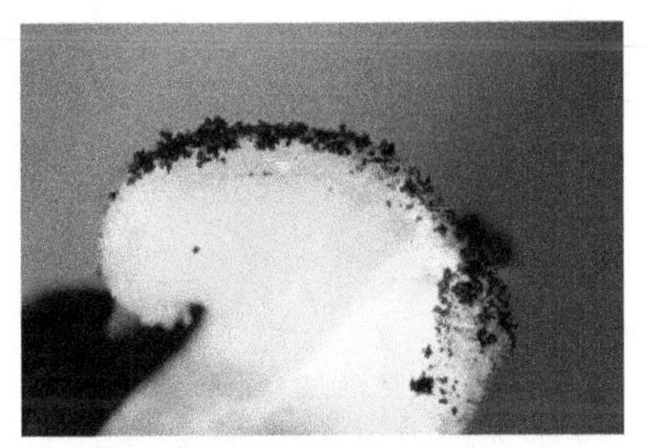

Stigma of a *Tulipa* species, with pollen

Corn stigma called “silk”.

The stigma, together with the style and ovary comprises the pistil, which in turn is part of the gynoecium or female reproductive organ of a plant. The stigma forms the distal portion of the style or stylodia. The stigma is composed of stigmatic papillae, the cells which are receptive to pollen. These may be restricted to the apex of the style or, especially in wind pollinated species, cover a wide surface.

The stigma receives pollen and it is on the stigma that the pollen grain germinates. Often sticky, the stigma is adapted in various ways to catch and trap pollen with various hairs, flaps, or sculpturings. The pollen may be captured from the air (wind-borne pollen, anemophily), from visiting insects or other animals (biotic pollination), or in rare cases from surrounding water (hydrophily). Stigmata can vary from long and slender to globe shaped to feathery.

Pollen is typically highly desiccated when it leaves an anther. Stigmata have been shown to assist in the rehydration of pollen and in promoting germination of the pollen tube. Stigmata also ensure proper adhesion of the correct species of pollen. Stigmata can play an active role in pollen discrimination and some self-incompatibility reactions, that reject pollen from the same or genetically similar plants, involve interaction between the stigma and the surface of the pollen grain.

Shape

The stigma is often split into lobes, e.g. trifid (three lobed), and may resemble the head of a pin (capitate), or come to a point (punctiform). The shape of the stigma may vary considerably:

Stigma shapes

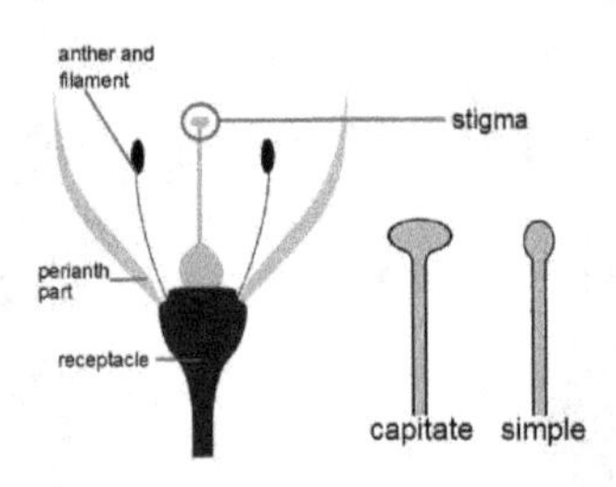

Capitate and simple

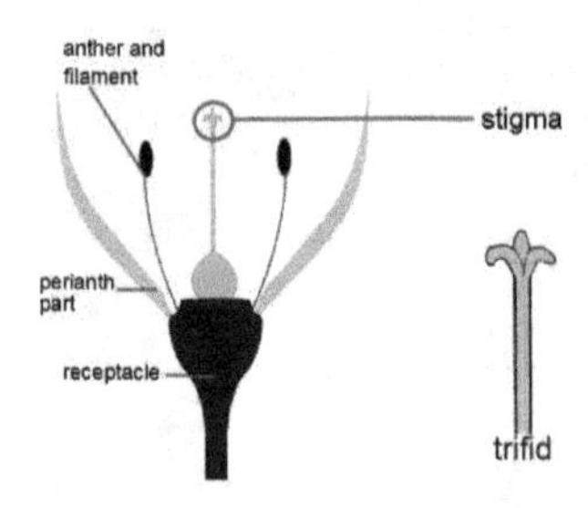

Trifid

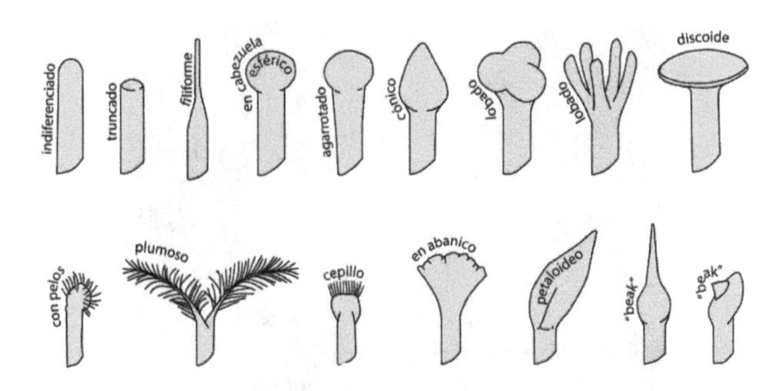

Style

Structure

The style is a narrow upward extension of the ovary, connecting it to the stigmatic papillae. It may be absent in some plants in the case the stigma is referred to as sessile. Styles are generally tube-like — either long or short. The style can be open (containing few or no cells in the central portion) with a central canal which may be filled with mucilage. Alternatively the style may be closed (densely packed with cells throughout). Most syncarpous monocots and some eudicots have open styles, while many syncarpous eudicots and grasses have closed (solid) styles containing specialised secretory transmitting tissue tissue, linking the stigma to the centre of the ovary. This forms a nutrient rich tract for pollen tube growth.

Where there are more than one carpels to the pistil, each may have a separate style-like stylodium, or share a common style. In Irises and others in the Iridaceae family, the style divides into three petal-like (petaloid), *style branches* (sometimes also referred to as stylodia), almost to the base of the style and is called tribrachiate. These are flaps of tissue, running from the perianth tube above the sepal. The stigma is a rim or edge on the underside of the branch, near the end lobes. Style branches also appear on *Dietes*, *Pardanthopsis* and most species of *Moraea*.

In *Crocuses*, there are three divided style branches, creating a tube. *Hesperantha* has a spreading style branch. Alternatively the style may be lobed rather than branched. *Gladiolus* has a bi-lobed style branch (bilobate). *Freesia*, *Lapeirousia*, *Romulea*, *Savannosiphon* and *Watsonia* have bifuracated (two branched) and recurved style branches.

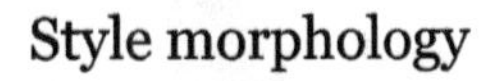
Style morphology

Iris versicolor showing three structures with two overlapping lips, an upper petaloid style branch and a lower tepal, enclosing a stamen

Iris missouriensis showing the pale blue style branch above the drooping petal

The feathery stigma of *Crocus sativus* has branches corresponding to three carpels

Attachment to the Ovary

Style position

Terminal (apical)

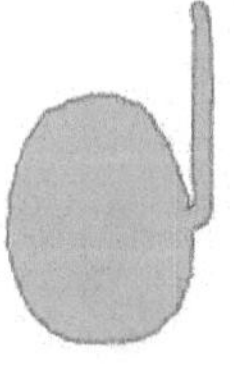

Lateral

Gynobasic

May be terminal (apical), subapical, lateral, gynobasic, or subgynobasic. Terminal (apical) style position refers to attachment at the apex of the ovary and is the commonest pattern. In the subapical pattern the style arises to the side slightly below the apex. a lateral style arises from the side of the ovary and is found in Rosaceae. The gynobasic style arises from the base of the ovary, or between the ovary lobes and is characteristic of Boraginaceae. Subgynobasic styles characterise *Allium*.

Pollination

Pollen tubes grow the length of the style to reach the ovules, and in some cases self-incompatibility reactions in the style prevent full growth of the pollen tubes. In some species, including *Gasteria* at least, the pollen tube is directed to the micropyle of the ovule by the style.

References

- Fritz, Robert E.; Simms, Ellen Louise (1992). Plant resistance to herbivores and pathogens: ecology, evolution, and genetics. Chicago: University of Chicago Press. p. 359. ISBN 978-0-226-26554-4.
- Rooting cuttings of tropical trees. London: Commonwealth Science Council. 1993. p. 9. ISBN 978-0-85092-394-0.
- Reiley, H. Edward; Shry, Carroll L. (2004). Introductory horticulture. Albany, NY: Delmar/Thomson Learning. p. 54. ISBN 978-0-7668-1567-4.
- Reynolds, Joan; Tampion, John (1983). Double flowers: a scientific study. London: [Published for the] Polytechnic of Central London Press [by] Pembridge Press. p. 41. ISBN 978-0-86206-004-6.
- Judd, W.S.; Campbell, C.S.; Kellogg, E.A.; Stevens, P.F. & Donoghue, M.J. (2007). Plant Systematics: A Phylogenetic Approach (3rd ed.). Sunderland, MA: Sinauer Associates, Inc. ISBN 978-0-87893-407-2.
- Gifford, E.M. & Foster, A.S. (1989). Morphology and Evolution of Vascular Plants (3rd ed.). New York: W.H. Freeman & Co. ISBN 978-0-7167-1946-5.
- Sattler, R. 1988. A dynamic multidimensional approach to floral morphology. In: Leins, P., Tucker, S. C. and Endress, P. (eds) Aspects of Floral Development. J. Cramer, Berlin, pp. 1-6. ISBN 3-443-50011-0
- William G. D'Arcy, Richard C. Keating (eds.) The Anther: Form, Function, and Phylogeny. Cambridge University Press, 1996 ISBN 0521480639, 9780521480635
- Chamberlain S.A; Rudgers J.A (2012). "How do plants balance multiple mutualists? Correlations among traits for attracting protective bodyguards and pollinators in cotton (Gossypium)". Evolutionary Ecology. 26: 65–77. doi:10.1007/s10682-011-9497-3.
- Physics.org (2012). University of Adelaide. "Flightless parrots, burrowing bats helped parasitic Hades flower". Date Retrieved August 2013.
- Beentje, Henk (2010). The Kew Plant Glossary. Richmond, Surrey: Royal Botanic Gardens, Kew. ISBN 978-1-84246-422-9., p. 10

- "Human Affection Altered Evolution of Flowers". Livescience.com. Archived from the original on 2008-05-16. Retrieved 2010-08-30.
- Prenner, Gernard (February 2010). "Floral formulae updated for routine inclusion in formal taxonomic descriptions". Taxon. 59 (1): 241–250.

Permissions

All chapters in this book are published with permission under the Creative Commons Attribution Share Alike License or equivalent. Every chapter published in this book has been scrutinized by our experts. Their significance has been extensively debated. The topics covered herein carry significant information for a comprehensive understanding. They may even be implemented as practical applications or may be referred to as a beginning point for further studies.

We would like to thank the editorial team for lending their expertise to make the book truly unique. They have played a crucial role in the development of this book. Without their invaluable contributions this book wouldn't have been possible. They have made vital efforts to compile up to date information on the varied aspects of this subject to make this book a valuable addition to the collection of many professionals and students.

This book was conceptualized with the vision of imparting up-to-date and integrated information in this field. To ensure the same, a matchless editorial board was set up. Every individual on the board went through rigorous rounds of assessment to prove their worth. After which they invested a large part of their time researching and compiling the most relevant data for our readers.

The editorial board has been involved in producing this book since its inception. They have spent rigorous hours researching and exploring the diverse topics which have resulted in the successful publishing of this book. They have passed on their knowledge of decades through this book. To expedite this challenging task, the publisher supported the team at every step. A small team of assistant editors was also appointed to further simplify the editing procedure and attain best results for the readers.

Apart from the editorial board, the designing team has also invested a significant amount of their time in understanding the subject and creating the most relevant covers. They scrutinized every image to scout for the most suitable representation of the subject and create an appropriate cover for the book.

The publishing team has been an ardent support to the editorial, designing and production team. Their endless efforts to recruit the best for this project, has resulted in the accomplishment of this book. They are a veteran in the field of academics and their pool of knowledge is as vast as their experience in printing. Their expertise and guidance has proved useful at every step. Their uncompromising quality standards have made this book an exceptional effort. Their encouragement from time to time has been an inspiration for everyone.

The publisher and the editorial board hope that this book will prove to be a valuable piece of knowledge for students, practitioners and scholars across the globe.

Index

A

Anemochory, 10, 58, 203

Anemophily, 160, 162, 171-173, 196, 228, 256

B

Ballistic Dispersal, 57

Barochory, 57, 203-205

Bat Pollination, 178-179, 199

Bee Pollination (melittophily), 197

Beetle Pollination (cantharophily), 199

Bird Pollination, 180, 199, 206-207

Butterfly Pollination (psychophily), 197

Buzz Pollination, 176, 197, 209-210

By Animals (zoochory), 11

By Water (hydrochory), 11

By Wind (anemochory), 10

C

Cleistogamy, 165, 181, 183-186

D

Dicot Germination, 35

Diseases, 2, 24, 62, 66, 89, 170

E

Edible Seeds, 17

Embryo, 1-9, 12-17, 22, 26-27, 29, 32-35, 52-55, 70, 116-117, 125, 127-129, 133-134, 139, 146, 149, 160, 194, 216, 218, 244-245

Embryo Nourishment, 9

Endosperm, 2-4, 6-8, 13, 17-18, 28, 30, 49, 91, 127, 129, 133-134, 146, 149, 160, 194, 218, 245

Entomophily, 160, 162-163, 173, 206, 228

Epigeal Lilies, 39-40

Etiolation, 23

F

Fruit Tree Pollination, 210-212

G

Gene Diversity, 63, 68-69

Genetic Diversity of Seed Orchard Crops, 68

Germination, 3-4, 8-16, 19, 21-23, 25-27, 29-49, 51-59, 61-62, 65-66, 70-71, 73, 76, 78, 87, 127, 134-135, 146, 161, 202-205, 213, 256

Germination of Bacteria, 37

Gymnosperms, 1-2, 5-7, 22, 30, 36, 53, 116, 118, 123-124, 126-136, 138-139, 143, 147, 159-162, 168, 172, 182, 194, 202, 217-218, 232-233

H

Holcoglossum Amesianum, 184

Hybrid Seed, 19

Hydrochory, 11, 58-59, 204

Hydrophily, 160, 162, 173, 196, 228, 256

Hypogeal Lilies, 40

I

Inducing Germination, 15

Institute of Plant Industry, 78

L

Lily Seed Germination Types, 39

M

Maturation, 5, 23, 52, 102, 128

Megagametophyte, 3-4, 125, 127-129, 162, 244

Monocot Germination, 35

Moth Pollination (phalaenophily), 198

Myrmecochory, 11, 57, 60-61, 205-206, 221

N

Nutrient Storage, 7

O

Open Pollination, 19, 66, 181-182

Orthodox Seed, 20-21

Ovule, 1-4, 7-9, 36, 122-123, 125-129, 132-133, 144-146, 160-162, 182-183, 186, 194, 217-218, 228, 240-241, 244-245, 259

P

Paphiopedilum Parishii, 184

Pests, 24, 62, 89, 102-103, 109, 111, 159, 191

Photomorphogenesis, 23, 28

Plant Propagation, 21

Plant-insect Pairings, 177

Plants Pollinated By Buzz Pollination, 209

Pollen Germination, 36, 161

Pollen Vectors, 166-167, 183, 195, 197, 200, 228

Pollination, 19, 36, 66, 68-69, 103, 111, 124, 127, 132, 134, 150, 160-187, 189-203, 205-207, 209-214, 216, 218-219,

224, 228-229, 231, 245, 249-250, 253, 256, 259
Pollination Syndrome, 172, 195, 206
Precocious Germination, 36

R
Recalcitrant Seed, 19
Repair of Dna Damage, 15, 185
Resting Spores, 36

S
Seed Bank, 52, 75-83, 85-88
Seed Coat, 1-2, 4-9, 13, 15, 23,
26-28, 32-34, 52-54, 122, 124, 127, 133, 146
Seed Contamination, 63-65
Seed Dispersal, 56-61, 201-206, 221
Seed Dispersal Syndrome, 201-203, 206
Seed Drill, 63, 70-73
Seed Enhancement, 63, 65
Seed Germinator, 51
Seed Orchard, 63, 67-70
Seed Paper, 63, 67
Seed Production, 2, 19, 58, 63, 66, 68-69, 158, 185, 202
Seed Saving, 65-66
Seed Testing, 63, 70
Seedling, 7-9, 11-12, 14, 22-27, 29-35, 37, 39, 41, 43, 45, 47, 49, 51, 53, 55, 57, 59, 61, 65, 68, 127, 129, 134, 150
Seedling Growth, 9, 23-24, 65
Self-pollinating Species, 184-185
Self-pollination, 36, 69, 164-166, 182-185, 200, 211, 231
Soil Seed Bank, 52, 75, 78-80, 88
Spore Germination, 36, 134-135
Sterile Seeds, 16
Submerged Pollination, 173
Svalbard Global Seed Vault, 75, 77-78, 80-82, 84, 88

T
The Structure of Plant-pollinator Networks, 171, 191
Transplanting, 24

W
Wasp Pollination, 197
Water Pollination (hydrophily), 196
Wind Pollination (anemophily), 196

Z
Zoochory, 11, 59, 61, 204, 206
Zoophily, 160, 163, 178

www.ingramcontent.com/pod-product-compliance
Lightning Source LLC
Jackson TN
JSHW052207130125
77033JS00004B/217

9781635492552